Branches

ALSO AVAILABLE FROM BLOOMSBURY

Branches

A Philosophy of Time, Event and Advent

Michel Serres

Translated by Randolph Burks

BLOOMSBURY ACADEMIC
LONDON • NEW YORK • OXFORD • NEW DELHI • SYDNEY

BLOOMSBURY ACADEMIC
Bloomsbury Publishing Plc
50 Bedford Square, London, WC1B 3DP, UK
1385 Broadway, New York, NY 10018, USA

BLOOMSBURY, BLOOMSBURY ACADEMIC and the Diana logo are
trademarks of Bloomsbury Publishing Plc

First published in 2004 in France as *Rameaux*, by Michel Serres © Editions
Le Pommier, 2004

English language translation © Bloomsbury Publishing Plc, 2020

Randolph Burks has asserted his right under the Copyright, Designs and
Patents Act, 1988, to be identified as Translator of this work.

Cover design by Charlotte Daniels
Cover image © Andreus K / Getty Images

A catalogue record for this book is available from the British Library.

A catalog record for this book is available from the Library of Congress.

ISBN: HB: 978-1-4742-9750-9
PB: 978-1-4742-9751-6
ePDF: 978-1-4742-9749-3
eBook: 978-1-4742-9752-3

Typeset by Deanta Global Publishing Services, Chennai, India

To find out more about our authors and books visit www.bloomsbury.com and
sign up for our newsletters.

To those who did me the honour
of attending my courses

Contents

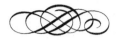

Preface

Today's world is screaming in pain because it is beginning its childbirth labour. At serious risk, we have to invent new relations between humans and the totality of what conditions life: the inert planet, the climate, living species, visible things and invisible things, sciences and technologies, the global community, morality and politics, education and health ... We are leaving our world for other worlds, possible ones, and will have to abandon a hundred passions, ideas, customs and norms brought about by our narrow historical duration. We are entering into an evolutionary branch.[1]

In previous ages, no knowledge had ever had to conceive or lead projects that were as vital: reinventing the universality of the individual, reconfiguring his habitat, weaving new relations. From having to rethink everything, philosophy changes span and sees its responsibility grow. Either a new human, a citizen of the world, will appear, or humanity will totter. We have to bring about peace between ourselves to safeguard the world and peace with the world in order to save ourselves.

System

Format-father

Anti-event protection system

Even though Venice, flourishing, was teeming with masts and merchandise, the Lagoon's shipowners and charterers feared bankruptcy. Why? Because, along with the rising numbers and greater distances, a thousand risks and wagers increased with the growth of their shipping exchanges; power on the open sea brought with it as many unforeseen events. Carefully laden, one ship sank, while another one, light, arrived in port safe and sound; the too dry season lost its harvest; in Constantinople, one correspondent underpaid, while the one in Marseilles died … To sell a coat in Padua, herds of sheep subject to epizootics had to be bred in the Balearic Islands; after shearing, the wool had to be carded in Narbonne; shipped by sea in calm or bad weather; stored in Genoa in rat-infested warehouses; woven in Rome; the garment had to be made in Turin, shipped again via roads crisscrossed with thieves … How can one not get lost in such an arborescent network of accidents and obstacles, interrupted many times over, knotted together elsewhere and differently for wheat, wine or spices? How can one master the flows of money circulating here and there, credits, long- and short-term debits, loans, dividends, exchanges between florins, ducats and other necessities, customs duties, octrois, taxes of every type, due in a space cut up by a hundred feudal systems – to say nothing of corrupt financiers, swindlers or other pirates, whose life even Cervantes shared?

Mediterranean commerce never would have unfurled, around the Renaissance, its wealthy profusion, crowned with immortal works of art; Venice never would have experienced such a splendour and this success if they hadn't concentrated, applied, even given rise to a thousand and one expertises as well as precise technologies and techniques: from the time-counter to the division of space into a network, from the fixing of units of every type to accounting …, measures intended to be prepared for any event.

Balance and balance sheet

The first books of this accounting described the whole of these operations in minute detail, line items and dates, transits and routes, costs and expenses, losses and profits, bills and projects. Their calculation aims at a general balance sheet that would put in order the firm of the person who maintains it and keeps it balanced, and would predict his future and prosperity. Just as, in his shop or the marketplace, the local merchant would hold a balance in his hand to weigh flour or salt and get a fair and accurate price and profit from this, so the shipowner and the banker, in their establishments, in which a hundred messenger boys and sailors hurried about, invented, so as to be able to launch operations to the four corners of the so-called inhabited world, invented, as I was saying, this accounting balance sheet, an abstract and generalized balance, as it were.

A refined instrument applied to a complex and fluctuating network, the balance sheet presupposed the gathering and support of other metric tools, already invented: balances, of course, to assess weights when preparing to cast off or unloading at the wharf, but also numbers and units (not only for these weights but also for lengths and volumes) suitable for contributing, through the series they form, to rules and accounts; clocks to mark time; a calendar that's precise and, if possible, standardized in order to bring the correspondents into agreement regarding dates, settlement dates and repayments; currencies whose value everyone agrees on, therefore a kind of Stock Market in which exchange rates would be debated; a common language, the lingua franca spoken by sailors and port travellers during those times; an arithmetic and processes of calculation, addition, division ...; natural logarithms appeared; processes for manufacturing paper, the cutting of books and page margins, the separation of words, the arrangement of chapters and paragraphs ... similar to those first musical scores equipped with notes named according to the first words of a hymn: *ut, re, mi, fa*[1] ...; printing soon after for letters; the rules of drafting for plans ...; perspective, in painting; a law allowing disputes to be settled ... How can one navigate, lastly, without instruments for measuring angles, without rules for calculating, without an astronomical model, without compasses or portulans, without a chart presupposing an exact network of space? These machines, these references contributed to a true mastery via this generalized metric.

This standardization of weights and measures, signs and acts made globalized exchanges possible among what was still called the inhabited world. In the space of this time, no economy of this size could have been built or could have functioned without these numbers, these accounts, not only without the measurements they permitted or presupposed, but above all without this generalized metrology. The economy resulted from this metrology at least as much as it conditioned it. Its newness lastly presupposed or accompanied the emergence of modern science at the same time and in the same places: Italy and the Flemish merchants, the respective countries of Gerolamo Cardano and Simon Stevin. Not, I repeat, that I believe that the economy determined the scientific discoveries, but the bringing together of a set of technologies and techniques of the same type, of a type I'm trying to name, conditioned the complex business exchanges around the Mediterranean, at the time of the Renaissance, and the emergence of the science we inherited.

Format: Homogeneous units, series, repetitions

I call this type format: accounting or countability formats exchanges, the way the *computer*, today, formats our information and globalizes it; let's likewise call the standardization of the set of these measurements formatting, whether the measurements be scientific, practical, cultural or even artistic, as in the case of musical scores for choirs and masterpiece paintings.[2] I take accounting [*comptabilité*] as my example because the word precedes the word that designates, in Anglicized Latin, our machines; by grouping the countable [*comptable*] and the measurable, both of them carry out a thousand formattings, presupposing elements, series, repetitions, homogeneity, measurements and agreements concerned with domains we instinctively think have no relation.

The Renaissance, as I've said, discovered the same characteristics in the musical staff, the writing of its notes, of its tempos and rhythms, as well as in the perspectivization of space, the clock counting of time, lastly in Galileo's experiments (the son of a musician) on heavy objects, their speeds and accelerations. Did the father, in opera, combine techniques of measurement brought together by the son in mechanics and astronomy?

Things and humans: Formation

While countability unites and names a thousand kinds of counting in this way, it also makes contracts necessary. How are we to reduce everything to countable equilibrium? *Com-putare*, comparing two or several things together, unites the preposition *cum* with the verb *putare*, to reckon or think, itself stemming from *putus*, clean and pure, used for the refining of gold or silver, but also for the cleanliness of people; Latin readily unites *purus* with *putus*. We can only compare two volumes or weights of gold if the metal doesn't contain any impurities; in this way we will be able to repeat an operation whenever we wish. The purity of the things allows us to reduce them to numbers. Extracting, abstracting: purifying every alloy conditions repetition, that is to say, abstraction, that is to say, lastly, the law. Thus we will be able to string together series, calculate ... count.

Here we find the true nature of metal defined; this principle of identity applies to the subject as well as to the object: no base metals in pure gold; no cheating in exchange. Purity of the benchmark, entirety of confidence, unsullied by any theft. The market requires erasing corruption. When pure, neither things nor humans lie. Contracts become possible. Formatting attains the individual and groups, culture and thought, the pedagogical education of humanity and the dogmatic Reformation of the Church.

Whether it's a matter of letters or numbers, of days or seasons ..., of printed books or Bibles ..., units of the same nature repeat themselves in series. Physical and mental, hardware and software, an entire universe becomes regulated and formatted. The movement historians later called humanism varies according to the same word: education [*formation*] of young people, new methods of teaching; available information, the Bible at everyone's disposal; transformation of minds, Reformation. A universal formatting of humanity and its world: Mercator's earth emerged from the format of his cylindrical projection; Copernicus's world came out of the heliocentric format; Vesalius's anatomies formatted cadavers, which Louvain had allowed to be cut up.

The definition of format

How, once again, are we to define format? That for which repetition is law; a generalized measure groups the set of these units. Was Western mastery born from this tremendous synthesis of metrics? This has been said.

Format concerns humans and things, nature and culture … as well as the event, its opposite. While we can quickly make out the power format procures, here are its drawbacks. Preserving measurable smoothness allows effectiveness of course, by eliminating all accident, but excludes the event and forbids the news.[3] Examples: television shows cut images up at a rapid rate, assess the intensity threshold of catastrophes, the number of invited guests, the character of the TV host and so on. No real news will ever penetrate this stiffness. The books published by universities require precise subjects, bibliographies and indexes, obligatory quotations, footnotes and so on. No invention can penetrate this rigidity. Formats succeed without inventing, imitate without innovating. Doing away with impurity planes down the unexpected and therefore newness. Will minds fall in this way into the countable? It therefore pleases me to think that in *In Praise of Folly*, Erasmus joined in Montaigne's dance, the laughter of the people riding in concert with the knight of the book with the sorrowful face and Rabelais's farces … amid a globalization of the economy and of culture, whose homogeneity was threatening the joyous celebrations of invention. From the formatted stem, a branch will be reborn.

A different genealogy of the computer

The Renaissance didn't organize these alignments starting from nothing: as far back as the Early Middle Ages, monasticism was following the canonic rules of Saints Bruno or Benedict, whose edicts set time schedules, the hours and the rite of Divine Office, the way the enclosure cut up space and the way the clothes made the monk. The latter, in addition, abided by canon law, whose name evokes the addition of two formats, and by the liturgical year. An old Church word (it appeared in French during the same era: 1584), the *computus* calculates the epact, or the interval between the rhythms of the Moon and those of the Sun, in order to base the ritual schedule, based on the sliding around of movable feasts, like Easter and Pentecost, on the fixed feasts, like Christmas or the Epiphany.

Obeying a strict schedule, for the year as well as for hours, Matins, Prime, Terce, Sext, Lauds, Vespers, Compline, roused from the cell by the bell, leaning over the lectern at study, aligned side by side in the refectory, always in rows and often silent ..., he who has never known boarding school in his youth, training ships later, lastly shift work in a factory or in a business ... is lacking in his education. These collectivities follow ancient rules similar to those that organized monasteries. The cloister, the schoolyard, the factory or the office, the stadium, the prison ... carve out a place in space, count time, determine the occupation of the days and the sequence of hours, clothe their monks, boarders, sailors, team members ... in uniforms. In brief, they format time, space and the actions of children and adults. This programme has varied little from the Middle Ages up to yesterday morning.

Pedagogy, production, training

Far be it for me to criticize the idea of the formats propriety today would call jails. Only a few years ago, entering to teach in La Santé, Paris's central prison, I was hardly surprised to see that its architecture reproduced down to almost the smallest details the architecture of the school where the boarders of my youth experienced a similar confinement in space and the rhythms of time; have I ever seen more attentive listeners? Punishment or labour? Forced labour for the sake of reform or education [*formation*]?

Shoeing horses or dehorning cattle, in the past, required that we immobilize them; they were tied up in a parallelepiped made out of wood beams joined by iron brackets; no farm was lacking this trave, in the format of three stakes.[4] An instrument of torture, we now read in books written by city people who have never seen one; but, without it, how were we to equip the animals and make them fit for working the fields? They resisted at first, not long; it all happened as though they understood their interest. An acrid odour emanated from the plough oxen, at the burnt horn, as well as from the white-hot nails meant to attach the horseshoes to the mass. Like these domestic animals, I went, bound, to the 'trave' of the boarding school, from which I exited shod with sciences and letters. At least I learned what iron collars and shackles to resist.

I shall continue my praise of format: today like yesterday, there can be no work without traves, without the format of such a framework. Of wood and iron. He who has a calling for writing will attain this if he enters the monastery. Going to sleep and waking at regular hours, leaning over the work for a constant duration, without missing a day, he will only align his paragraphs by likewise aligning his limbs and moods, his space and his time, the whole of his existence, dedicated. The will to regulation adapts to the autonomous regulations of the body, heat or heart, and regulates them to its norm. Anchorite, writer, it's the same battle. The West has forgotten the meaning of the Latin *devotio*: the passionate sacrifice of the flesh. Do you want to write? Face this death. Without format, whose forms concern the worker as well as his work, there can be no production. How does an athlete attain what he calls being on form? Through a training requiring him to follow a rule and to become a monk, like the writer. It's a matter here of a necessary condition, which ensures, at least, a well-made work, an honourable race, a mediocre rank among professionals. For genius, the necessary condition has not yet been found. The classification of the sciences and the disciplines, of articles and theses, footnotes, indexes and bibliographies, honest quotations and humility in debate …, university constraints discipline research and thought. Conform to the iron collar and shackles of the formatting ...; obey the format-father, who reigns, invisible and absent, over absolute knowledge. But if you want to invent, take risks; abandon the format. Even if it means dying, become son. Great works unite format and invention, iron discipline and freedom: father and son.

Leibniz: The mathematics of the father and the singularities of the son

An example: pre-established harmony, in Leibniz, presupposes that God, the King of power and Father of mercy, Creator of all eternity, programmed the cosmos and humans; this harmony regulates the monads in a gigantic monastery; the cloisters I was talking about are organized and function like a reduced model of this universal harmony, whose formatting generalizes the entire enterprise from just now to the universe. It has taken me my entire life to understand that this philosophy, systematized according to mathematical models (I wrote this thesis in the past and dedicated it to my father, a peasant and bargeman), referred to this haughty figure that is the Father, to his laws without any exceptions, formatted for all eternity.

Up until the individual and relational *Singularités* described by Christiane Frémont persuaded me that events, non-reducible to these formats, trace out, faced with this Father, the Son's place; that the Christianity of the incarnation completes monotheism, in the strict sense; that, faced with formal sciences, a subtle logic puts events and newnesses in place. In short, that starting from the trunk, universal and necessary, of the format, a thousand contingent suckers explode in Leibniz. Extremely rare in thinkers, this double entry, ramified, now passes to my eyes as a good criterion for authentic philosophy as well as for the philosophy our times long for.

But, before getting to that, let's see how many formattings – paternal – have seen the light of day in the West.

Five moments of formatting in the West

1. Around the Aegean Sea, the sixth and fifth centuries BC saw the invention of alphabetic writing through the cutting up of phonemes into vowels and consonants; the invention of money, another form of impression and division into elements (our word 'pecuniary' remembers that *pecus* used to signify the herd to be bartered); followed by the Platonists, the Pythagoreans grouped their arithmetic geometry around the theory of proportionality; via his homothetic theorem, Thales rigorously described the prior form of the pyramids, under whose weight and volume Egypt attempted to format unformattable death. From science to languages and exchanges, the formattings cover the same span as that of the Renaissance.

Whether it's a question of models of the world with the Presocratics, of ideas in the style of Plato, of Aristotelian logic and mechanics ... all the way up to the incipit of the Gospel according to St. John ..., the Greek word *logos* doesn't so much signify discourse, speech or knowledge, the conventional translations, as it writes $a/b = c/d$. Quasi algebraic, this formula designates a measure as well as a balance assessed on a set of scales, a spatial invariant of form across variations of size, a justice and a harmony, and therefore organizes the exactness of measurements the way it tiers the musical scale by the vibration of strings; it guarantees the arrangements of contractual apportioning, and therefore politics and morality as well as relations with God: we humans have the same relation to Christ, St. John said, as the relation that unites Christ, the Son, to God, his Father. In the format of this formula, the Greek miracle is summed up. In four letters and three signs, this *logos* traced the first Western format, universal enough not to distinguish any discipline or specialized field, from metaphysics to religion, from algebra to geometry, from law and justice to politics, from poetry and song to the organization of harmonic forms. With Hippocrates and his school, it brought about a new human body, which sculpture imitated by means of an omnipresent golden ratio; with Thales, for whom everything comes from water …, with Empedocles, who roots the world in the four elements, it brought about a new earth, measured in latitude by the gnomon's shadow by Pytheas of Massalia, and according to the meridian by Eratosthenes; starting from Anaximander, the same geometric ratio or relation brought about a new universe. As with the Renaissance, a different human inhabited a new measured cosmos.

It is related that Athens invented democracy; by means of this lie, it eliminated women, foreigners, metics and slaves from politics. Yet, by means of this *logos* or ratio, Thales showed the shadow of a body to be proportional to the shadow of the Great Pyramid; the pharaoh's power, lying under this lapidation, becomes compatible with the weakness of whoever may be standing there. Better than institutions, this *logos* says equality.

A different synthesis

Why does Plato in the *Statesman* devote the greater part of his preparatory efforts to questions, pointless at first glance, focusing on dichotomy and the art of measurement? Why does he evoke, as we do with the Grand Narrative, the rotations of the world and its temporal phases? Why does he pass technologies in review, the way we just visited Venice; why, lastly, does the art of the weaver become the figuration of kingship ... if not because politics demands a general regrouping of every format? Plato, it is said, invented the Ideas or the Forms. This signifies that this Greek time found the art of formatting in everything humans fabricated in culture or encountered in the natural world: counting, spelling, reducing matter to elements, cutting with a knife, articulating, dividing, measuring, aligning, intertwining series ... Arts and trades, mathematics and logic, education and orthography, world and living things, herds and groups, politics lastly, the supreme art ..., the perceptible world and social collectives abide by the forms of the intelligible world ... beneath the paternal shadow of the royal weaver.

2. Latin genius gave this general metric in the forms of law and administration, drew land and sea routes, set up an army aligned into legions themselves arranged into cohorts ... More concrete than formal, more cultural than natural, more jurist than physicist, Rome formatted humans more than things, governance more than technology. In social terms, it attained universality just as much as Greece.

3. I won't go back over the Renaissance, in which the flowering of standardized elements I just spoke about appeared. Once again, the invention of printing resides in format: the type case, into which Gutenberg put the forged letters, allowed one to move from written letters to other written letters, manufactured, manipulable, substitutable, and vice versa: the first hardware formatting of software.

4. With the metric system, around the end of the eighteenth and the beginning of the nineteenth centuries, one of the first attempts at global formatting appeared, universal because transcultural. Relating to the dimensions of the planet and astronomical rhythms, the decimal system moved little by little from science, manifestly universal, to a usage that's almost everywhere accepted, beyond the traditional units, whose diversity made their translation as well as crossing borders by humans and things difficult, even if it took lots of time for the new units to assert themselves locally. Linnaeus's language, which is suitable for designating living species, and the language created by Lavoisier for chemical elements and compounds ... prepared this enterprise.

During the same revolutionary era, a new calendar attempted to format time by relating it to the weather – wind, rain, snow, fruit and harvests … – so as to free it from religious references, which had limited its use. A hopeless enterprise, formatting the regular by means of the irregular! The fact that, unlike the metric system, these new namings had quickly failed doesn't prevent me from sometimes repeating, with delight, the stanzas of the ecological litany: *Prairial, Messidor, Vendémiaire, Nivôse* … Later, the positivist calendar, intercultural, united, to the contrary, sciences, arts and religions by enumerating fathers in preference to a nature Auguste Comte rejected as metaphysical; but the Law of Three Stages and the hierarchical classification of the sciences had already formatted history and knowledge.

A new earth, from which weights and measures emerged. A new universe: co-author of the metric system, Laplace formatted the solar system according to Newton's law, thenceforth, thanks to him, as universal as the languages of chemistry and natural history. From the sciences to law and politics, the breadth of coverage we had noted in the preceding episodes is found again. Should the French Revolution be taken, on the other hand, to be an unforeseen and new event, or did it carry out projects prepared by the Age of Enlightenment and the *Encyclopedia*? Did it free from the old format, or did it, conversely, impose the formats that preceded?

To wrap up this era, I would readily justify the importance sometimes given to Kant by the extension of these formattings to subjectivity (knowledge, morals and judgement): *a priori* forms of sensibility, the schematism, concepts of the understanding, regulative ideas, the categorical imperative regulating formal morality, definitions of the sublime and the beautiful ... format the subject the way metric units do the world. The fact that the naiveté of the enterprise makes people today laugh or worry doesn't prevent its author from having regulated the inward.

Format, information, recording medium

5. I would never have attempted, so imprudently, such an incomplete retrospective if we hadn't witnessed today a new and similar attempt, a universal one. Let's redefine format: it concerns the size of a sheet of paper or, now, the dimension of any recording medium. This usual sense prepares, at least formally, the universal metrology in which we are living. Rare are the actions, practical gestures, exchanges or relations ... in which such metrics don't now intervene. Without these metrics, there can be no medicine, no commerce or control, no law or society, no police, no morality or politics, no thought, science or religion. Formatted in this way, the encyclopaedia added up on the internet and the Grand Narrative's temporal integral replace, today, the grand treatises, from Euclid to Laplace ..., whose publications cadence the above moments.

For computer and information science [*l'informatique*] generalizes the old attempts once again since the format it proposes defines the rules to be followed for the size and arrangement of information on a recording medium. Used independently of their content, these last two notions have of course just allowed us to revisit history by uniting domains that were distinguished or hierarchized without justification. Whether it has to do with graffiti on walls, theorems on parchment, notes on a score, poems by email ..., representations of every type, drawings or messages, lies in love letters or falsehoods by telephone ..., molecular chains in a gene, a crystal or a cell, even atoms in a molecule ..., we find there information deposited on a recording medium, itself invariant in function across variations of texture.

How do we deposit this information there? The general idea of code generalizes, in its turn, figures and elements previously enumerated: letters, notes, numbers, the chemical elements of genetics, the folding of molecules ... Bits and pixels don't take content, matter or meaning, the real or the virtual into account: there are as many information units in Pythagoras as in Verlaine, in insults as in the Law of the Twelve Tables, the memory of a computer, the DNA of an organism, a chemical reaction ..., as many similar pixels in the *Mona Lisa* as on the screen of a mobile phone or the photograph of a galaxy.

Since information is measured in proportion to its rarity, its value disappears into the minuscule in comparison to ordinary energies; as for format, whether ordinary or global, its enframing or depositing on a recording medium doesn't add any information. More general than those of the preceding metrologies, these new concepts, formal, of format, medium and coding promise a more powerful mastery of the world, whether inert or living, of cognition and practices, by occupying things and classes of knowledge. Of course, we exchange information with each other, whether as groups or as individuals; we would die from not doing so; but we decipher it, coded in living things, molecules and atoms. Right where yesterday we still saw exchanges of energy, we now spot transfers of information. So we are beginning to understand why certain molecules form while others disappear, as though the laws of evolution had entered into the inert world. Matter and life contain the repetitive format and the rare information expressing newness ... – the global drawing of the stem and the branch.

Hence the necessity of a new synthesis, following the one by the *Statesman* or by Leibniz's system. Philosophy would miss our time if it didn't seek, through such synthesis, to reconstruct, like the two others, the cognitive, the objective, the collective. How many of our institutions resemble those stars whose apparent sparkling we still see but which we know to have been dead for ages? How many resemble museums! A thousand fossils clutter our cities. This broad synthesis would allow us to reconstruct everything.

Hominescence celebrated the greatest discovery of the century; I am retranslating it: within the very heart of matter, in the arrangement of atoms in order to form molecules or the arrangement of particles in atoms lies information. Everything in the world, ourselves included, receives it, preserves it, transmits it. Every element formerly called material or hardware says *logos* or software. Consequently, the act of formatting information descends into the elements. Alive and thoughtful, we are no different from them; the history of humans is grafted onto the Grand Narrative of the world. There can be no universal formatting that would be more universal. Thus coding, format and information enter into metaphysics, one in which the recording medium replaces substance. This new synthesis binds the universe and cultures via a natural contract.

The Grand Formatter

Internal or external, immanent or transcendent, does the format come from the world itself, from elsewhere or from us? We sometimes form an image of it: simple, symbolic, carnal, moving ... Here are some of its figurines, its pixels: a father runs his family; a president governs; a legislator says the law; a strategist steers battles; a scientist masters an expertise; a doctor heals; a teacher teaches a language; an orator gives language voice; an architect designs buildings; a masterworker perfects his masterwork; a banker invests in businesses balanced by accountants; a priest or pastor preaches; an adman pollutes space and time with brands; a sage lives morality; a saint, a genius, a hero, a champion set examples ... A thousand textbooks give names, variable according to language and ideology, to these admirable and doubtful titles.

In arranging these elementary vignettes in a large format, whether the oval of a painting or the rectangle of a screen, they are collected together and take on meaning, at a high scale, under a single and same heading; the figure of the father adds up this multiplicity. This is the synoptic integral of these shadows, the Grand Formatter, however varied the domains in which these rules are applied may be. The preceding historical moments define the eras of the Father ... Plato's royal weaver, Leibniz's God ... Have we ever abandoned this image?

The father, today?

Laws that are more regular than the laws invented by past ages to organize or enslave collectives, to understand and dominate the world, to save or subjugate souls, today format the Grand Narrative of the inert and the living, before cultures are born, therefore in the absence of human intention. Consequently, what face of the father do the laws of physics, of genetic programming, the global grasp of the narrative itself ... project onto the giant screen of the universe and the gigantic duration of the cosmos? Will we find there the set of the figurines whose list I just gave? Outside space and in the eternity of time, is the God of philosophers and scientists – all-powerful, omniscient and creator, just and merciful ... – returning for a handy synopsis of the format of all formats?

Does immanence regulate itself in silence, or does it apply a transcendent word? Can thought do without the integrative function, whether personal or not, of the understanding-sum of the eternal truths evoked by Leibniz? Can it unfurl itself without a global system whose equilibrium and closure guarantees the faithfulness to the real of our local thoughts? Do our pieces of knowledge resort to the regulative function of an absolute knowledge? Even the Age of Enlightenment lived in the grip of this summational form, literally preformationist, a perennial spring and reference for laws, for reason, for every formatting, physical and cultural.

Those who seek to kill the father consider this format to be brutal, not without some appearance of argument. In fact, the

quantity of violence required to impose some format can easily
be assessed. The strategist, the legislator, the king … dispose
of armies; God surrounds himself with legions of archangels
wielding swords of fire; the financier creates wealth and poverty
… In the face of injustices, whoever doesn't revolt lacks courage.
Some reasons, colder and demonstrative, on the other hand,
dissolve the very idea of system and preformation. Do formats
become set up contingently? I shall return to this question.

The tyrant

Let's turn to more contemporary violences. Does today's anxiety
come from the fact that we feel ourselves to be bound up by a
tremendously insistent formatting? We no longer lift a finger
without paying a tax for the privilege; myriads of images invade
our representations and format them in their turn; we no longer
feel a single desire without advertising having already aroused
their brands in our automaton-souls; world and environment,
acts, objects, emotions and opinions imprison us with closely set
bars. We are no longer even going to have children blindly … So
we link all formatting with necessity, necessity with enslavement,
and enslavement with death. Freedom, what has become of your
old victories? Death, have you just triumphed? The figure of the
father slips into that of the tyrant. Fundamentalisms reflect its
sinister remains.

Our behaviours consequently reverse the old attitudes
concerning the unexpected and unpredictable, events that
distressed our ancestors so much that the realizations of the

formats reassured them instead, inasmuch as these realizations followed the predictions of scientific laws. The Age of Enlightenment sheds light on this idea.

Foresight and prevision

Foresight there counted as a virtue of the father of the family, with finite sight and limited horizon, whereas the other Father, eternal, all-powerful and creator, was considered to be the only one to know everything, by means of a rational prevision[5] coupled with a merciful providence; this triple variation on vision distinguished the two fathers, the eternal one from the temporal one. As soon as Newtonian laws allowed predicting an eclipse as well as the impact of an artillery shell, prevision, formerly reserved for God, passed to the temporal father; God alone enjoys this prevision, Voltaire said, whereas foresight, random, remains human; we are now attaining not only foresight-virtue, Maupertuis responded, but also prevision-knowledge. A blasphemy to the ears of Voltaire, this advance heralded progress in the eyes of Maupertuis.

Inhabiting and living in a world given over to the vagaries of events-accidents, contingent and dangerous, these neoclassical thinkers felt themselves to be partly freed from their risks by the sciences and technologies, which create formats. The way Prometheus stole fire from heaven, we stole prevision from God. The Enlightenment therefore taught us to respect laws, not only the laws whose spirit Montesquieu described or the ones whose legislator Rousseau named and whose contract he defined, but the laws of physics, mechanics and natural history, laws that are regular and allow previsions. Starting from Newton, we knew how to govern the forces of the world. Even if it had a tendency to overturn all or part of theology, the Enlightenment continued to revere the Father; mastering prevision, the Enlightenment set itself up in his place.

We traverse this Enlightenment perspective twice. We inhabit and live in a world already so regulated by technosciences, so protected, smooth and previsional that we love the event and the new instead. We experience prevision less as a benefit than as a loss of freedom. We fear to find ourselves lacking precisely what used to make our ancestors afraid: contingency. In addition, as I shall have to say, today we are entering not so much into divine prevision as into its creation; so, refusing the risks, we reject the invention of man-made objects whose effects we cannot predict. We fear the sciences and technologies doubly: when, continuing the Enlightenment, they eradicate the vagaries of contingency; when, continuing the old vagaries, they eradicate our security. We prefer to trust the accidents of nature, provided, of course, that they remain gentle.

An age of the Son?

We are separated from the Enlightenment less through belief in God or atheism than through the difference, far more decisive, between preformation and unprevisionability: through another fear of contingency. We now create things whose behaviour exceeds our previsions. Creation becomes separated from preformation. Contingency returns everywhere, including in knowledge said to be hard; for example, prevision, mechanical, becomes fissured by chaos theory. The old triple law of the father, prevision and providence, only leaves us still with foresight.

Another example: we reason and have to decide by means of large numbers and singularities, with margins of error. When, united, the journalist and the politician demand a settled decision, the scientist today answers with a statistical estimation, that is to say, with a doubt and a proportion. When the previous two demand zero risk, the scientist denies this possibility. So the figure of the father splits: the twofold power of the media and the state perpetuate, before the public, the figure of the father, whose understanding contains preformed certainties so as to still allow belief in definitive and closed truths, in a tribunal of cases and accusation, in a world system, in a rational sequence of history ..., whereas the scientist quits these formats so as to adopt a lower profile in which knowledge changes status and inaugurates a time I characterize as the age of the Son. He now says that there is always a margin of error, and therefore of risk; I think, therefore I doubt; he who doesn't doubt doesn't think, but plays at the ancient figure of the father.

Hence the new counter-list of sons: the scientist hesitates in his expertise; the doctor is charged before the courts; the sage wavers between impossible choices; the legislator codifies questions he no longer has any mastery of; the father negotiates the running of the family with the mother and children, and the tyrant flees his country, ruled as a democracy ...; the strategist worries about no longer being able to kill children ...; *Homo faber* and the theologian lastly understand that no one ever masters his creations ... The filial age comes.

Sciences-daughters

As a result, the various revolts against the father, against the stupid and repetitive habits, against the preformation without freedom to act ..., against the sciences and technologies, whose laws bear the same name as the laws dictated by a domestic and public king ..., recruit their militants from the same camp as their adversaries, who lament the loss of references without seeing that contingency exceeds the format; that preformation has never taken place; that the possible exceeds the previsional; that large numbers always add an accursed portion to a regulation that only norms a majority; that freedom comes within easy reach. When law abounds, detail superabounds. Aspirin heals, but its spread is not exempt from accidents; no absolute certainty guarantees that some person will not die from it instead of being relieved by it; reason, science, law ... the father ... prescribe it, and it will relieve the pain of millions of people; but no one now forgets the inevitable margin of lethal risks. The event returns across the format the way a flood filters in trickles over and under and through a dike.

A new knowledge-son, which the mathematician himself has practised ever since he proved that no formal system closes over itself, unloads the bearded legislator, the lion grown old … of their self-importance, their self-assurance, their certainties. Proof, reason … format leave a residue, in every place, both natural and cultural. The real is scattered around the rational. The concrete surpasses the abstract. Singular instances exceed the rule. The flesh outstrips biochemistry. A person and his singularity draw varied landscapes such as are taught to the doctor and the magistrate by the particular, in its turn. Hands seek to hold a water that always exceeds their grasp … This disequilibrium between format and the informal, law and the multiplicities that exceed it, this deviation, this existence … move the world, living things, history, cultures and the sciences …, cause, here and there, a thousand arborescent branches to surge forth.

Let's continue on … from the simplest – we don't know the law of distribution for the prime numbers themselves – to the most complex, which produces this dialogue: Einstein, 'God does not play dice'; Niels Bohr, 'Are you claiming to tell God what he must do?' At twenty-four, Werner Heisenberg exceeded, with the profusion of nuclear nanodetail, the figure of the father, with his flowing white mane, who was dictating his law to the world. These brief words, exchanged between the triumphant relativity and the beginnings of quantum mechanics, which was running counter to intuition, reproduce, almost word for word, those that Bossuet addressed to Leibniz, who had just described how God had created the world: 'Do you have privileged access to the divine design?' Singular and inconstant, tattered clouds often veil the Sun in such a way that everything always seems new under its father-figured single law. Chaotic, weather conditions can prevent observation of astronomical laws. Such laws themselves, so regular in Newton and Laplace, have exploded with chaos since Poincaré or the galactic clouds of astrophysics. Landscaped, circumstances return into knowledge. Crowds of sons exceed the father.

Kill the father? Not so fast! The existence of an element of death in mastery remains an unfathomable mystery of production. Should it be necessary to revolt against the tyrant, that's fine. Avoid, however, replacing him once victory is achieved the way Napoleon sat on the throne of kings and Stalin took the czar's palace. The existence, conversely, in the format, of an irreducible motor of production, which I have praised and will praise, remains an equally unfathomable mystery. We need it to live, think or produce. Yet, irresistibly, it leads us to death. Seeking to kill the father to free oneself from him consists in falling back into the same law of death. A question remains: How does one escape this fate?

Erasing the contingent event in the name of rational law seems to me to be just as unreasonable on the part of the ancient father as doing away with the law to the benefit of the abundant real on the part of the new son. Might we bring them into agreement? When the theory of branes and superstrings attempts to reconcile the relativity of the first with the quantum mechanics of the second, I dream that mathematicians are settling a family argument.

The Grand Narrative and history

Before human freedom, human madness and human will format their own customs, the Grand Narrative abides by two formats, both properly universal: the laws of physics and the genetic code. The second one, whose origins are unknown to us, continually mutates and produces, contingently, thousands upon thousands of diverse living species, taking the environmental filter into account; the combinatorics that shapes its variations resembles a lottery more than a set of decrees. As for physical constants, which accompany the laws of physics, they are the hallmark of both the real and its contingency; for if they had other values, it would be a matter of another world. Their product erects Planck's wall behind which lies, upstream, a beginning over which we have no hold, but which, downstream, opens the constitution of this particular world; for, produced differently, another beginning would have marked out another path, whose direction would have ended up at another world. Consequently, all formats, those of the Grand Narrative and not only those we decide on, necessary of course with regard with their application, become contingent with regard to their beginning. Born contingent, they become necessary. Like masterpieces of art.

The Grand Narrative then, in its entirety, obeys the following modalities: rare and full of information, a contingent event tends, with duration, towards a necessary law, a format without information; in its development, possibilities, fluctuating around it, disappear, mercilessly pruned by impossibility; sometimes one of them emerges, contingent again, surges up and, resistant to the impossibilities, becomes, in its turn, necessary ... True of the inert, this succession of branches applies to evolution and to my existence as well as to cultural, scientific or artistic productions, in sum to the Grand Narrative. Epigenesis, then, winning out over preformation, changes the figure of the format-father under the influence of the event-son. How is knowledge to be conceived? What the social sciences called history becomes science, and what the hard sciences named science becomes history. System and Narrative exchange their values.

Paleoanthropology describes us as sons of Humanity. This re-rooting universalizes this title to every culture. The definition of humanity is continually being constructed; we are permanently working towards hominization, towards the birth of sons who would be more human. We find ourselves fathers of Humanity: even concretely, in our laboratories. To kill the father or to love him, this is a question of epistemology, of cognitive science, of ethics; it concerns the continuation of the Grand Narrative that we are fabricating. Once again, this question arises in the same way in the hard and the social sciences, in the arts and religions, for knowledge and the real.

Science-daughter

Father-son: Deduction, induction

A painting, again: in *The School of Athens*, visible in the Vatican's *Stanze*, Raphael painted Plato, the father of philosophers and scientists, standing in glory to the left of Aristotle, right at the top of the steps of a portico, with the *Timaeus* in hand: the Quattrocento still read the beginning of the world in this dialogue. The Renaissance, later, on the contrary, turned away from a demiurge preforming the universe by means of mathematical models, abandoned, likewise, the prime mover from Aristotle's *Metaphysics*, to obligate itself to experimentation and humbly subject it to the decision of the real. Did it master the real, after having obeyed it in this way, as Bacon had dictated? Not quite: the Renaissance only falsified theory, as would be said later. Modern science quits mathematics in the Greek style, in which deduction commands, in favour of a more inductive method, which isn't guaranteed success. Abandoning mastery for subjection, modern science erases the haughty image of the father Raphael painted in glory.

A few decades ago, Alexandre Kojève intuited that the dogma of the Incarnation had made, in the Renaissance West, the invention of this experimental science possible, a science turned towards the world as such and no longer deduced from abstract theories. The union of geometry and experimentation achieved by mathematical physics continued in some way, he said, the union of the divine and the human, of another world and this one. In his eyes, Christian theology constituted the cultural condition for this cognitive innovation. This Renaissance forgets the father and his deductive model; by itself, mathematics cannot predict what formula will explain a given phenomenon. A given equation, on the contrary, is born from experiment. Kojève was right: the new knowledge promotes the image of the son. From the top of Greek knowledge, the father deduces the world; the son submits to its reality.

Decenterings

What does the world show after this era? Fallen from its ancient royal situation of being centre, the planet Earth, after Copernicus, became marginal, a servant, the daughter of the Sun, which, in its turn, would later quit the central throne to become one star among others, the daughter of a galaxy, itself the sister of a plurality of other galaxies, nieces of dust, descendants of light. Dislodged from the central pole, the Earth turns around a star, which soon becomes shifted from the centre of the Milky Way, itself plunged in a universe in which all places, in the end, are equivalent. All the centres abandon the centre; all the kings abandon the throne: homogeneous and isotropic, the universe nowhere lets any place be seen where the father can sit and reign in glory and order movements. Astronomy and astrophysics, over four centuries, carried out as many successive decenterings that continually unhooked from the father-position. Laplace's planetary system, deterministic in accordance with its paternal demon, becomes unstable in Poincaré, in which unpredictable chaos already appears, while chaos theory announces that the successors or sons, by turning around, can know their ancestor perfectly, but that this latter cannot predict or preform his succession. The father sat on the throne at the centre of the world, the possessor of strength and reason, the prime mover; each father, by turns, through revolution or desire, sought or took this polar place, held it for a time up until he quit it, lastly up until it disappeared. Now there is no longer any centre or even any notable place in a universal space without any privileged site. The big bang itself doesn't enjoy any central position: the birth of the Universe took place in every one of its points. No more father, not even a first one. The universe expands in the image of the son.

Even better, *The School of Athens* ordered the space of knowledge around two centres, dominating the steps, Plato and Aristotle, the fathers of philosophers and scientists. The world revolved for a time around the two foci of an ellipse; spiritual and temporal, power displays its splendours around two thrones, the pope and the emperor, media and politics. In the parallelism of these three images, knowledge becomes confused with dominance.

No, knowledge doesn't function like power, as the image of the father would have us believe. Reason doesn't always and everywhere prove someone right. Necessary, certainly, but it isn't sufficient. The world doesn't arise from reason and reason alone; knowledge arises from reason, from humans and from the world. He who claims to hold knowledge loses it. Desirous of such an appropriation, we yield to the desire for mastery, to a social Darwinism, wrongly generalized from evolution, in which the dominant males, whether elephants or presidents, try to vanquish and kill, not to know. The cognitive continuously requires a humble self-abasement. The father orders, the son knows. I like the philosophers crouching on the lower steps.

Scientists-sons

Not only Diogenes, the dog, or Pyrrho, the sceptic … Gödel stated the incompleteness of formal systems; Heisenberg declared the indeterminism of quantum mechanics; general mechanics ended up at chaos. From the heart of the most rigorous axiomatic system to the most refined equipment of experimentation, the history of our science continually augments the same retreating movement from every position of certainty and puts them in doubt. Discovery-daughter: the universe, contingent, evolves contingently, functions according to contingent laws and constants having contingent values. The world and our science quit necessity. We no longer lay claim to full mastery, not in fact, not in principle and not by deduction. A similar contingency shapes the evolution of living things; amid the evolution of species, *sapiens sapiens* loses its situation of being the human-source, of being the centre, I was going to say of being the trunk, to take on the situation of being a branch. The subject of this knowledge-daughter takes the place of the son. A book-son, *Branches* plunges into contingency.

Not only Epicurus, Lucretius and the *clinamen* … Look at, during the French Revolution, the glory of the father Lazare Carnot, a classical mechanist, the organizer of victory, the president of the Committee of Public Safety, and the poverty of the son Sadi Carnot, dying crazy in the Charenton hospital and the inventor of the new thermodynamics. The one killed in the name of a deadly past; the other, dreaming of the new, constructed the future. The history of science repeats this canonic couple over and over again. Descartes ended his days while staying with Queen Christina; Pascal defended Port-Royal, razed by Louis XIV, Stalin's bewigged ancestor. After the *Discourse on Method*, the one invented a geometry that didn't go beyond Greek geometry; the other discovered new ones and the path to future algorithms. Cartesian past, Pascalian future. Abel, the well-named algebraist, whose discoveries were lost by Cauchy in the Academy; Gallois, the author of modern algebra, dead in a duel at the age of twenty; Mendel, alone and unread in the depths of his monastery; Boltzmann, driven to suicide in Trieste; Semmelweis, condemned by scientific Europe for having saved the lives of the pregnant; Wegener, ridiculed for his intuition of tectonic plate theory …: scientists-sons whose views were ahead of their times abound and incarnate the adventure of knowledge, intuition, heroic wandering, invention … Crushed by Cuvier, Geoffrey Saint-Hilaire and his unity of plan in organic composition heralded homeoboxes; Pouchet, defeated by Pasteur, had prepared the ground for the prebiotic soup …; in the conflicts opposing men or schools, often look for the defeated; many Nobel Prize winners saw their projects rejected by ad hoc committees presided over by male and female mandarins. Truth comes on dove's feet; the silent thief arrives unexpectedly in the night. Neither the bird of peace nor Hermes the pickpocket figures among the fathers in *The School of Athens*. Living science was born daughter. But who, better than her, follows the law?

Father and Son

How many ideas that were so sure they were taken to be dogmas have disappeared from knowledge? How many ideas that were reputed to be absurd founded knowledge in reason? Even Newton's idea, one of the great successes of modern science, at first appeared to be crazy, and for good reason: how can bodies, without any magic, attract each other at a distance? From irrational numbers, imaginary numbers and other circular points at infinity all the way to plate theory and large molecules of the protein type, which no one believed in, how many impossible intuitions took reason by surprise and replaced it? Called the big bang as a term of abuse and scorn (the stupid big bang, Fred Hoyle said), it became true as soon as the background radiation of the universe was revealed. How many people believed, at their beginnings, in the impossible intuitions of quantum mechanics? Bohr's response is known: the truth doesn't establish itself through its own content but because the preceding generation goes into retirement. This is an excellent definition of science and its history: the father continually goes into retirement in it; the son holds a hesitant and temporary place.

Whether rationalist or not, epistemologists and philosophers took and take a position with regard to atheism, creationism, preformationism …, in short, with regard to God the Father. Few people meditated on the Son, except Pascal, sometimes and not particularly well, since he opposed the God of Abraham, Isaac and Jacob to the God of philosophers and scientists; except Leibniz, as we have seen; except Nietzsche, who, after the camel and the lion, sang of the child; except, above all, Saint Paul and, after him, Montaigne. 'I think', says the omniscient father, laying claim to an absolute knowledge. 'What do I know?', hesitates the son. 'We know', affirm classical rationalism, logicism, formalism, hypercriticism itself … 'For how long?', replies the historian of science, following Montaigne's example.

Possession or contract?

Bacon, I repeat, advised obeying nature in order to command it; going one further, Descartes recommended becoming its master and possessor. As advances continue, more and more things in fact depend on us. Today we understand that we will never attain definitive mastery, for we increasingly depend on the very things that formerly and recently depended on us. In the face of antibiotics, microbes that are penicillin resistant are returning; we are paying for our waste of energy with water and air pollution. Our formats are contracting debts. A mastery acquired in one place brings into play, endlessly, a new obedience in another place. The entirety of our dominations gives rise to constraints that impel us from behind. The dominant one, a killer, commits suicide at the limit of his practice; for lack of new victims, microbes die at the end of the epidemic; so it is for predators, for lack of prey. We must renounce the father's dream. Symbiosis, obligatory, opens on to a natural contract. The father dictates the law; the son negotiates agreements. Knowledge takes on a contract form.

Juxtaposed to these scientific and epistemological discoveries is the weakening of the *pater familias*' kingly rights, which stem from anthropological memories. Will a politics be born from new contracts? Let's dream that a republic-daughter will accompany the knowledge-daughter.

Birth and newness

The son is born, new, but will not reign. There can be no science without birth, newness, perpetual invention, refreshing the landscape … without ramifications. Should the father, from his retirement, return one evening, he wouldn't recognize anything anymore – everything would have changed. If he wanted to stay with dogma in order to maintain himself, he would pillage the future ..., or he would adapt and then become son again: everything new is fine. He is reborn.

Ulysses returns to Ithaca: he puts everyone there to death because everything appears in a new way around Penelope, where the suitors were crowding. The father can't stand invention or change, especially not one where his wife, who he has been cheating on for some time, enjoys seeing herself surrounded by admirers. He pierces them with his arrows. And, in killing, becomes centre again, potentate, petty king of his laughable island ... and so ceases to know, stops his discoveries, takes his retirement, him, the ancient inventor of islands, the explorer of known and unknown lands, the lover of Circe, of Nausicaa, the prisoner of father-Cyclops, the dinner companion of king Alcinous ... He was reborn each time he was shipwrecked. This is the scientist: disembarks, washes ashore, wakes up, stands up, exhausted, is born, on the beach, before a new landscape, where he encounters, playing ball on the beach with her companions, Nausicaa, who takes him with her to her father-king's home. Alcinous gives a banquet where Ulysses, in the position of a shipwrecked sailor, of a stranger, of an impoverished starveling, in short, in the situation of an adoptive son, recounts at the father's table the encyclopaedia of his discoveries. To my knowledge, Ulysses doesn't kill Nausicaa's father. He honours his table and his daughter.

This old picture book, tied to the Trojan War, already recounts the murderous connection between knowledge and power. Among our violent drives, I hope that knowledge will abandon the position of strength held by the father; this latter kills all better because he knows, drunk with a reason transfixed by ideology, bringing about the conviction of the less educated. Peace can in part come from a knowledge in the position of son, from a culture-daughter, peaceable, who would not kill her mother.

Mother, black widow

In high school, we learned to revere Andromache, the pious and faithful widow, the loving mother, devoted to the memory of her dead husband, whose image she finds in her son Astyanax, hostage like her with their victorious enemies. Her constancy shines forth in a hallucinatory narrative in which the fall of Troy and the murders of that night blaze: may that night, she says, remain eternally. 'Should I forget?', she repeats. No, I live in the immortality of memory and reject the course of history: I shall no longer live, no longer love; I shall consult my husband's voice over his tomb; I shall only be addressing him when speaking to others …; lastly, I shall kill myself right after my second wedding … The dead require nothing but death.

A tragedy of remembrance, whose actors Racine designated by the name of son and daughter (of Helen, Agamemnon or Achilles), *Andromache* relates the misery of the second generation. What could be more terrifying for a child than to hear his mother tell him: when I hold you in my arms, Astyanax, I am embracing Hector, your dead father …, than to obligate him to carry in his body an adult in the form of a corpse? The mother-widow teaches the sons and daughters of the finished war to do nothing but cry or die again from war, the way their parents died from it; not at the point of weapons, but from the deadly disease of remembering.

At the end of the play, Andromache brings off the feat of marrying her enemy and becoming a widow, once again, so as to reign, sovereign, over the dead and the madness of those around her, a double-headed widow, a double queen, of the Greeks and of the Trojans, a black culprit of murders. We called her spider's motionlessness perseverance; we turned her time-disease, indifferent to every change, into a virtue, faithfulness, but also into a knowledge, history. Its deadly cost must be calculated: around and because of Andromache, tragic murders increase despair and frenzy. She alone will survive and reign, a black widow, an abusive mother, saturated with the death instinct, a repetitive spider in the centre of the web, devouring with her old teeth those fine and strong young people, who only asked to live, to love, to hope in the future. Let's suppose that conversely she had let the dead bury the dead, agreed to live, followed the course of the living present, run to new loves, conceived a project …, then life would have taken despair's place. The ancients had seen things correctly: Mnemosyne, the Memory mother, engendered the Muses, among them Tragedy, terrible.

Who can deny it? Without history, we would return to being animals. So an obligation to remember is necessary, a tie that holds us to language and no doubt to consciousness; but a duty to have projects is especially necessary. More difficult than the first, the second requires imagination, discernment, a sense of the present, of anticipation, a will to survive for the sake of following the heading decided on, enthusiasm, courage … transcendent virtues in comparison to repetition, itself falling towards the instinct of death.

History and tradition sustain us, of course, but they only find their meanings through the rereading a sustained future makes of them. We don't die so much from enemies or obstacles as from the lack of descendants or production, in the bed of immobile anamnesis's infinite detail. Without firm intention, the past falls into forgetfulness; a collective without resolve no longer knows how to write its history; without invention or living contemporary works, a culture is dying. Memory digs our grave and, on this closed foundation, projects build our abode.

Christmas and Palm Sunday [*Rameaux*]

The Son is born: son-event, event-son. The branches [*rameaux*] never stop shooting up; the branches of the year, of the season, of time, those that bifurcate, those of unexpected history, fragile, slender, bristling, trembling in the wind. Every morning, in his office, in his laboratory …, the son tells himself a story he didn't know the day before. He tests, tries, risks and doesn't repeat or copy. Of the arborescence, he deals with the green foliage, the grafts, the suckers, the points of the twigs. Being reborn, he observes things being born. Nature, *natura*, that which is going to be born. He signs the natural contract every morning with a partner who is newborn entirely new. *Naturus*, he who is going to be born. *Deus sive naturus*, God or the Messiah to be born.

The scientist resembles the travelling Ulysses; before he returns to Ithaca, a murderer – he tries to escape the black widow. Or he resembles Christopher Columbus ... a new world announces itself. The explorer leaves Ithaca, Venice or Spain, the quay, port, capital and palace, in which the nobles, divided into pressure groups, fight one another to decide things they know nothing about. Departing, he left the social and political theatre, the representations of power, the unreal drug of parasite relations. He is born every morning to the adventure of the real, itself always fluctuating. The Son is born, not in a palace, but in the straw of a stable; goes away, leaves his family, wanders, a prodigal son, teaches nonetheless to pray to the Father ..., and when, before dying condemned, he returns to Jerusalem, he appears perched on a donkey, amid palm branches.

Knowledge

Knowledge is different from what is said about it: approximate, disquieted, ignorant and naive, obedient to experiment, running in proximity to error, always put to the test, changing and patient, light and mobile, often lost, always ardent, impassioned to the point of madness, resigned to strange intuitions and to never savouring victory. Authentic discovery is ahead of its time to the point that no one understands or hears it, like, just now, the night burglar. The public reserves its understanding for formats it already hears or understands, therefore for repetition, rarely for invention. Glory consecrates the repetitive fathers. Wandering and wayfaring, knowledge – not true knowledge, but a genuine knowledge – abandons power for knowledge, society for objects, glory for intuitive flashes, short life for the long term, this world for the other one, politics for curiosity; it undertakes three travels.

A trip around the world first; there is only one world, and amid the mountains and seas, plains and glaciers, deserts and shores, seagulls and whales, tarantulas and kangaroos, algae and trees, knowledge, local and meticulous, incomplete and lacunary, gets lost, however long and carefully it may have travelled a thousand lands. Get up, grab your staff and walk, leave your country, throw away your sandals, run across space, know, if you lose your way, that in the open space of landscapes, beauty expands. We shall have to talk about the plurality of worlds.

A trip through society next; there is only one humanity; knowledge gets lost there, from bargemen to the ultra rich, from the poor to functionaries, from farmers to presidents, from Australians to African Bushmen, from the Aborigines to the Japanese, from the Chinese to the Bolivians …, a knowledge that's local, friendly, incomplete and lacunary, however long and attentively it may have encountered its fellows. Get up, leave culture, class and language, have the courage for alterity; deep down in humans lies goodness. Melt your soul into so many belongingnesses that a new culture will not frighten you.

A trip through the sciences lastly; there is only one Grand Narrative, temporally gigantic, and knowledge gets lost there among the sciences and their history, algebra and biochemistry, cosmology and botany, geography and logics …, Greek geometry and Newtonian chemistry, classical infinitesimal calculus and the mathematics said to be modern …, a knowledge that's always local and meticulous, incomplete and lacunary, however precisely and rigorously it may have examined a thousand pieces of knowledge.

But while this wandering knowledge, a daughter, lacks completeness and authority, it does provide the incomparable joy of connecting together, here, there or elsewhere, the maps of the three trips: this glacier with the Moroccan guide and the physics of the globe; this Australian tree with the Aborigine and biochemistry; the River Garonne with my bargeman father, ignorant of fluid mechanics, and with my rock-breaker brother, ignorant of crystallography. Knowledge follows the path of the crest – dangerous, thrilling and often broken – of concordance.

What a quasi-religious jubilation, linking together science, humans and the world. Oh knowledge, daughter of joy.

Wandering or exodus

The *Odyssey* describes the voyages of father-Ulysses, the master of ruses, a king before the abominable Trojan War, restored upon his return among the corpses; Exodus relates the voyage of Moses, Aaron and Joshua, guides, just men, fathers of the chosen people, victors over the Pharaoh and the Sun; in the Latin *Aeneid*, Anchises still crushes Aeneas's spine with his weight.

We wander in knowledge like in the *Telemachy*, accompanied by chance mentors who are only right for a time but who we never cease to respect. An adoptive son, Jesus wandered in Galilee, Judea, Samaria …; who knows the address of his house since he lived homeless? Another traveller, Saint Paul widened his wanderings to the oecumene, sleeping outside or with a chance host, in circumstances he himself recounts, hunger, thirst, cold, wild beasts and chanced-upon bandits, fatigues, illnesses …, always goaded by a thorn buried in his flesh. In *exodus* rather than *methodus*, our science has no abode: nor stable assurance, nor definitive certitude, nor closed axiomatic system, nor sure prediction, without hearth or home. It haunts tent or tabernacle, like the Hebrews in the desert. Our knowledge extends this wandering to the universe, a universe equipped with the arborescent voyage of its Grand Narrative, followed by the astray unpredictability of the human adventure. Through its global contingency, through local and temporary hesitations, knowledge constructs its weakness from filial wanderings that lead it to humility; no more triumphal roads to royal certainty. The universal space of the world and knowledge no longer takes one back to the island or the Eternal City, nor to the valley where milk and honey flow.

Metaphysics and metanomics

Long after the Presocratics had invented physics, the Greek librarians who had to categorize Aristotle's works called the books following the ones in which the master had written his *Physics* metaphysics; they thus designated their place on the shelf. Later, modern scientists had the generous wisdom to accept that a metaphysics existed, a genuine discipline beyond their knowledge, as exception or complement, a strange knowledge whose title the Latin languages preserved in Greek. Since it deals with non-falsifiable questions, this metaphysics ensures the sciences an exactness announced by Popper's criterion. Why did the sciences become falsifiable? Because they had the properly filial humility to leave undecided questions to an associate metaphysics, which would thus ensure a protective roof – others would say a convenient garbage can. As soon as physics adopted the status of science-daughter, it accepted, as higher than it, a sort of science-mother, non-falsifiable; a father, the metaphysicist is right without limits. Science merits this title if and only if science is not always and everywhere right: in other words, when the scientist becomes son.

To measure the advantage of having a metaphysics above oneself with exactness, it is enough to compare this division of tasks with the state of the social sciences. We who, in an often dishonest way, use the term sciences for the disciplines that study cultures and collectivities have had neither the generosity nor the wisdom to accept that phenomena exist that exceed their limits and their claims. And therefore we haven't wanted to invent a supercultural domain that could have given said sciences a metanomic extension – a discipline that would, for its part as well, have dealt with their non-falsifiable questions.

But, by that yardstick, almost all this knowledge would have been transported to this new extension. On balance, authentic sciences become daughters; should the other scientists remain in the state of fathers, they will miss knowledge.

Reason [*raison*] in the sense of *arraisonnement* – the forcible boarding and inspection of one vessel by another, stronger, one – issues from the father; reason in the sense of knowledge issues from the son. Everything dies from power: the Persian and the Roman Empires ... So much for history; and this, now, is in order of evolution: species disappear; the human animal is subject to the temptation of mastery; happily, humbleness came to it on the blessed day it knew: and could then know death. We find ourselves devoted to humility in our active behaviour as well as in contemplation. Philosophy wouldn't be worth one minute of trouble without this gentleness.

Learning, inventing

From having learned, we know. A truth, a piece of information were found amid the internet's ocean, in a tradition, by way of an interlocutor, at a chance person's home ..., and we received it through education, communication, hearsay or effort. So we join an expert group, participate in a community, even in an institution, whether school, research unit, library or data bank.

Thus acquired, knowledge extends towards three dimensions: by the first one, cognitive, I know some theorem; by the second one, collective, I am part of those who know it and who, sometimes, put it to good use. I readily call the third of these dimensions stony, inasmuch as this information doesn't transform me any more than it does a rock I hold in my hand, which I can transmit, of course, but can also forget or let drop. I know but don't comprehend.[1] I can teach this theorem; it can thus be spread, but I take said knowledge, objective like this stone, to be as cold and dead as it is. In the ad hoc discipline, we do indeed speak of dead information.

And all of a sudden, by means of a process I can only shed light on by comparing it to digestion, which transforms a piece of bread into active biochemical elements in my body, or to pregnancy, which transforms an oocyte into a fetus, I make this theorem mine. The time it takes is indeterminate: a fraction of a second or decades; how many times, two or three decades having passed, have I violently felt, from my thighs to my thorax, that this digestion, that this conception were finishing their work and that I was entering into a true comprehension of what I had merely known. Quickly therefore or little by little, the theorem passes into my head, my eyes, my original perception of my landscape, my genitals even, my active life; I walk in its space, place my hands and feet according to its measure, inhabit and caress its forms in such a way that I recreate it, reinvent it from its foundations; this objective changes into subjective. I no longer know it – I feel it, live it, comprehend. On this pebble and other ones, I build my bones, my body and its habitat. So I can generalize said theorem, enter into the geometry modelled by it ..., but, yet again, I see and repeat structures already produced by others before me ..., and the digestion, gestation

and incorporation recommence ... up to the point that I haunt
these structures like a house that's become mine, a lodging I fix
up and repaint, whose walls I make plumb again, whose garden
I cultivate, whose blueprints I redraw, which I try to reconstruct
by making it bigger ..., this habitat describing, metaphorically,
the very site of my bodily life ..., flesh and dwelling becoming
subjective and objective at the same time ...; this is the time of
inventing. I find myself close to childbirth, to that externalization
I shall later call exo-Darwinian.

Whether lightning fast or slow, this passage from knowledge
to comprehension and from learning to invention is experienced
and lived by anyone who devotes his time and existence to a
meditative labour. But what should we call this experience I
have just called digestive or impregnated, this transformation
of an outside object into personal and carnal subject, this
incorporation, if not transubstantiation, formerly a miracle,
an experience that's inward, evident, vital, and which changes
bread and wine into body and blood? I can testify to the fact that
daily labour wouldn't amount to very much if this thaumaturgic
mutation of the pebble-theorem into bone and muscle didn't
take place continuously and this very morning. Consequently,
knowledge becomes unforgettable, both for me, since it has to
do with my blood, and for a few others, since knowledge can,
afterwards, become objectivized, setting sail from the body as
new inventions.

So mysterious that its method can't be given by anyone,
the art of inventing commences in this metamorphosis: the
objective transforms into subjective; knowledge changes

into comprehension; bread transubstantiates into body, and, once again, body into bread. For it sometimes happens, next, that this very body becomes externalized as a new object through this kind of exo-Darwinism: the subjective produces something of the objective and of the collective. The group then re-appropriates the things thus externalized. I promise to talk about this latter process in connection with technological objects.

The *escient* of consciousness

To give this living comprehension another name, I would like, poetically, to reincorporate into my language the old noun *escient*, whose use today is reduced to a few locutions used absentmindedly.[2] The root of the word, the Low Latin ablative absolute *meo sciente*, me knowing that, had already completely united the subject of knowledge and the object. *À bon escient* describes this way my body lives with the object of its research for a long while, through it, with it and in it, so close to it that, inseminated in me, this bread becomes my body, develops in it without me, then, if I invent, it can find itself condemned by peer pressure, but has a chance, tomorrow, of defeating this death. Comprehension implies a metamorphic and shifting life; life implies this productive comprehension that's victorious over death. Eroded from below, it is said, by the unconscious, might consciousness, above, draw its own fibre out by the *escient* or living comprehension, before invention, the ultimate fibre of sur-vival or super-life, quasi supernatural? We experience that the *escient* exists and can believe credible witnesses about it since they give tangible signs of it, their discoveries, unlike the underlying boxes, hypothetical and black.

On balance, this (cognitive, objective, collective) changes into subjective. Dead information becomes living, resurrects. Newness appears in the cognitive body. Everything changes: the world and me, me seeing a new world and the world seen by me, new. There is no comprehension except for this birth, this advent. All the rest reduces to transmissions of stones from pocket to pocket, reduces to communications of information without any change, empty zero-sum games, dormant dead-sum exchanges. Genuine knowledge changes the body and the speech of the one who receives it, who gives it, who becomes transformed and transforms the bodies of others through his invention burning like a tongue of fire sent down over their heads.

This cognitive change of death into life – announcement, birth, deliverance from death … – is called Annunciation, Nativity, Resurrection or Pentecost by an ancient tradition. Its carnal experience encounters this kerygma, without any other mystery than this subtle transformation of bones and of the world carried out by the Word, than this imminent childbirth. Thinkers earn their living fighting against dead information, against the ice and numbness of things, of self and of others. In thus fighting against the rocks and the dead, from humble failures to painful defeats, they encounter the image of this Son, a humble carnal model of overcome failure: the Resurrected One. Good news and the art of inventing, as a pair, celebrate the era of the Son.

Communion of saints

Before invention, the objective becomes subjective. As soon as invention appears, the subject gives birth to an object a numerous group can recognize. As though, by dint of transiting, knowledge, up till then subjective flesh, then became again, via externalization, objective and collective. After the invention, the group again exchanges flesh become bread again. I incorporate knowledge into myself; invention makes it set sail from my body; an institution becomes incarnated around this new object. Famous moments in which the history of knowledge concentrates social forces and produces marvels.

I dream of the Solvay Conferences, during which the science of the twentieth century was built (relativity, quantum mechanics, information theory ...) through lightning fast transmission of refined and rapid inventions between inventors. And of the beginnings of the Bourbaki school, in which Chicago neighboured Nancy ..., of that post-war period in which biochemists travelled, constructing the genetic code by exchanging regarding the phage ... Of course, we only comprehend what we incorporate into ourselves; of course, we only invent on condition of this externalization. But who can guarantee that this genuine knowledge will in fact become a true knowledge, that this subjective body will be able to correspond, in its turn, faithfully, to something objective, to a new stone, soon exchangeable between ourselves, rather than to the previous one, if not the gathering of those who, bodily, entered together into this genuine knowledge? Without this collective, there is no truth, at least temporary truth; without this church, there is no salvation.

A cheerful parallel between the history of science and that of the church has continually shone forth before my eyes my entire professional life long: the same universality of a simple and difficult language in the case of mathematics, easy and complex in the case of parables; a similar contingent faith in a transcendent reality; the same fluctuations from a closed and semi-dead society under cardinals or mandarins to an open collective, light-hearted and dolorous, welcoming the mysticism of saints gone astray and of sequestered discoverers; the same dynamic in invention and conversion; the same dogmatic inflexibilities, upstream, the same incomprehension of innovation, even bifurcations, downstream, bifurcations sometimes called paradigm shifts; the same tribunals, the same fires, the same condemnations of heresy, as a curb, the same posthumous and hagiographic rehabilitations, the same type of education spreading a thousand tics for generations … debates, hatreds, similar fervours … Despite these myriads of faults, the fact remains that even the purist mystic, even the most brilliant inventor … can't isolate themselves under pain of a paranoiac idiolect. So without these two chapels, there is no salvation. A return to the format-father.

Who doesn't dream of a community speaking every language, beneath the fire of inventions shooting forth from all sides, so that Greeks, Levantines, Romans, Scythians, Jews and Galatians … understand everything that is said, each in their language, because they understand the real through their bodies? The Last

Supper, the festival of the Spirit, Pentecost, today's sciences. Who today attains, without any pretension, universalities of language, of object, of history, of community ..., a peaceful globalization, if not this knowledge? The collective animal Plato called, out of derision and disgust, the Great Beast, and Hobbes, out of fear, Leviathan ... transforms into communion of saints. It produces inventors in return. Filled with reason ..., we need, nowadays, sons of course, but above all we need saints.

But the historical and social tripod of the saint, the genius and the hero wobbles, rickety. The latter two, lifted on high, hoist themselves up there with great effort, while the first one, anonymous, doesn't seek any niche; only his counterfeit perches there. The only great humans are the saints, absent from the list of great humans.

Is this a naive view when war rages everywhere?

The end of war during the era of the Son

A return to the father, I said. What are we to do about the fury I just talked about? Listen to his cry: 'That he die!' ...[3] The old Horatius offers the life of his son to the country by calling on the Curiatii sons to dispatch him properly. Trumpeted out, these words cost him nothing and yield him glory on the stage. Who today, without being base, can call this cry sublime, this cry demanding an ignoble execution, advised by Roman law, this sacrifice of a child? *Hominescence* defined war as a contract signed between two fathers for the children of the one to be willing to kill the children of the other: to murder sons. Neither our presidents nor our generals fight, to my knowledge, but send their children to be executed. The fathers sacrifice the sons on the altar of the collective beast.

How should we define war? As a threefold limitation on violence: first, a historical limitation, in which a time (quasi cyclical) binds war (productive of the institution of the state) to the government (the institutor of war); a juridical limitation, next, since the declaration (prior), the law of nations (concomitant) and an armistice or a treaty (duly signed at the end) contain, through law, the misdeeds of vengeance, which could never stop; a ritual limitation, lastly, since, in putting the sons to death, the fathers behave as sacrificers. Politics, law and rite limited violence by, in return, rendering it productive of these archaic institutions. Through the state, the law and the sacred, war limits the total number of dead that violence in its free state would relentlessly produce all the way to possible eradication, a threat we formerly and recently held before our blind eyes, but our eyes today are clear in the face of this possible horizon.

'That he die!' ... I hear here, as for me, a persistence of antiquity, the bitter remains of the era of the Father: Greek sage, legislator, patrician and Latin *pater familias* ... The tragedy takes us back to this barbarous age, that of Agamemnon, the killer of his daughter Iphigenia, and of Andromache, the black widow. Fathers govern, legislate, sacrifice. They doubtless sacrifice to dictate law and laws. We haven't pulled ourselves out of this sacrificial age, even though the age of the Son has long struck: Abraham held back the knife over Isaac's throat; Jesus Christ died on the Cross but resurrected and lastly sits to the right of the Father, both of them having become peaceful. The religion of the Son, Christianity in principle ought to have inaugurated that era, which I finally see coming, for time and history have worn out our old juridical, political and ritual instrument for limiting violence, war, become ineffective, interminable, prohibitively costly and counterproductive: wiser than the decision-makers, who remain fundamentalists, global opinion – let's hail the recent birth of this universal subject – vilifies the victor more than the defeated, ranked among the victims. The age of the victim or era of the Son. Today, we are living through the death of war.

So a violence without known limit inescapably explodes since it is without law. So another barrier, a new one, is built, juridical once again, bound precisely to knowledge: a drift towards the objective. Today, environmental law is concerned about future generations. Strange and innovative, this respect for sons is the counterpart of the end of war, by sending the limitations on the violence of subjects to objects, of collectives to the world. The natural contract, the sole guarantee of peace? Further on, this book will renew its terms.

The adoptive son

Jewish, Greek and Latin, Saint Paul united in his single person three of the ancient formats from which the West was born. A pious Pharisee, born in Tarsus into a family of the diaspora, educated in Jerusalem beside Gamaliel, he respected Mosaic Law and continually cited the Torah, Psalms and prophets learnedly. We imagine that he knew Greek philosophy, at least through Philo, for, in that language he wrote, spoke and whose authors he sometimes mentioned, he occasionally happened to say that he admired its wisdom and feared its reason. A Roman citizen like his father, he prided himself on this rank; he knew the law since, condemned, he appealed to the tribunals of the Empire. No one has better described this synthesis than Stanislas Breton.

Saint Paul doesn't merely symbolize the cultural crossbreeding that was taking place around the Mediterranean during the *pax romana* among sailors, port merchants or a few rare persons of letters, but above all he achieves the total human constructed at that time by the law, *logos* and administration, three formats forged in the fires of Hebraic monotheism, Hellenic reason and Roman law, stemming respectively from rite in the temple, harmony in the *cosmos* and the city in the Empire. This triple belonging to an organized society, to a proportioned world, to an all-powerful God, promotes excellent behaviour. So three grandiose views forge Saint Paul's body, life and thought: the projection of eternity onto the time of history; *logos*, the measurement of the world and the language spoken on the sea and in the inhabited world as well as by rigorous knowledge; citizenship, spread by imperial politics. Ritual, rational and juridical, three rules sculpt the universe of antiquity: uniting three perfect norms, they define Saul's era, the name he bore before his conversion. Who hasn't known such human masterpieces, hasn't admired them, hasn't feared their fundamentalism?

Triply formatted in this way, Saint Paul, newly named, suddenly left the trinity of his belongingnesses, travelled the inhabited world, invented the era to come and, in so doing, confronted three failures: his coreligionists persecuted him; gathered at the Areopagus of Athens, the Greek philosophers mocked his jabber; doubtless, Rome judged and executed him. In him and through him, the stem of the best and most lasting of what the Indo-European and Semitic traditions constructed bifurcated; in him was incarnated and through him was grafted the news he announced; in him, the branch of a new creature surged up. The old formats in fact presupposed that he belong to three communities: the new human doesn't identify with any of these communities so as to create an original one. Which?

Belongingness and identity

In a previous work, I wrote: my identity does not reduce to my belongingnesses. So don't call me old, male or a writer; instead place me in some subset grouping respectively age, sex or profession. Beyond these implications, who am I? Myself. All the rest, including what administration compels me to write on my so-called identity card, designates groups to which I belong. If you confuse belongingness and identity, you are making a logical error, a grave or a minor one, depending; but you are risking a murderous mistake, racism, which consists, precisely, in reducing a person to one of his collectives. I was unaware, in the book in which I discovered it, that we owe this distinction, so important I love to take it up again, to Saint Paul: both in theory (because he set it forth) and in his life (since the good news he announced breaks with the old formats, all three bound to collectives).

'There is neither Jew nor Greek,' he said, 'neither slave nor free man, neither male nor female' (Gal. 3.28). Taken from Joel, this sentence only mentions classes, sexes, languages or nations …, in sum, collectivities; it signifies that there is no longer any belonging in the sense of just now and that this disappearance leaves room for identity, $I = I$: 'by the grace of God, I am what I am' (1 Cor. 15.10). The 'new creature' remains: *I*, the adoptive son of God through faith in Jesus Christ; *I*, with faith and without any work by which there would be cause to glorify oneself; *I*, empty, poor and null: universal.

Who am I? I am *I*, that is all. Abandoning the subset definition of belongingness ($x \in A$), the principle of identity is stated here, not in the formal way, $a = a$, as with Aristotle, but regarding an individual singularity, whose just anyone, even minuscule character Saint Paul often emphasized. I shall return to this emptiness. Better yet, this principle defines this singularity, not arbitrarily in any way, but by a free gift of God. Transcendence had accorded, in its mercy, election to a group; it now gives identity in the singular.

The first quotation refers to Greek, Hebraic and Latin communities, to social classes and sexual roles; the second one refers to the initial event to which the *self* owes its emergence: the incarnation of Jesus Christ, death and resurrection. These two short sentences therefore distinguish belongingnesses and identity for the first time. Identity tears itself away from belongingnesses. The News becomes detached from their formats.

By the sinful flesh, from which only faith delivers us, Saint Paul doesn't merely designate the body, needs and passions, but its being plunged in a collective whose fusion heat we love to feel, whose laws we love to be subjected to and whose reactive aggression we love to share. His Epistles thereby designate what I have recently called the libido of belongingness. We commit the majority of the sins of the flesh according to a mimetic impetus, out of peer pressure and in the blind enthusiasm of national, tribal, family … cohesion, out of corporatism or mafia. Who has the courage of the *I*? *We* commit them more often than *I* commit them, so much does sin concern the *we*, that is to say, the law,

and not the personal *I*, which delivers us from it. When Saint
Paul 'released us from the Law', he first and foremost liberated
our own identities from this collective bond.[1]

The newness of the *I*

Did this *I* ever exist in the eras preceding the Epistles of Saint
Paul? The citizens of Athens, democrats, that is to say, occupied
with carefully distinguishing their elite team from slaves, metics,
women and other barbarians, without work or profession,
occupied themselves with the affairs of their city; by set rites,
sacrificed to their eponymous goddess Athena; sometimes made
war against Sparta or the Persians: therefore organized, honoured
and defended their *we*. All of them together condemned those
who observed, objectively, the stars, and Socrates, who advised,
subjectively, to know oneself. The collective excludes objective
and subjective. How many Greek philosophers said I? These
political animals – I am quoting Aristotle – readily excluded the
object as well as the subject. Their measures, norms and formats
originated in belongingness.

Ever since its covenant, the chosen people has likewise turned
towards its law, respected it, honoured it, taught its children
its holy history, fought, when necessary, against the Philistines
or the Samaritans, expelled the εθνη [ethnē, gentiles] from its
temple. The *we* became reality in its contract of chosenness with
God, alone in saying the principle of identity: 'I am who or what
I am.' As far as I know, Rome, in its immortal writing of the law,

didn't designate, with these laws, any other categories than *paters familias*, senators, tribunes of the plebs … citizens …, all of them representatives of a group. There were no more persons in Rome than in Greece.

So at the opening of the first century of our era, the idea and behaviour of belongingness was covering the Mediterranean. Greek culture taught one belongingness, political and cosmic at the same time; Israel's tradition passed on a second one, holy; Rome's did so for a third one, juridical. The entire inhabited world practised another one, economic and social, which separated slaves from humans said to be free and born as such; dominant males lastly claimed that nature inscribed a last one, sexual, in the body. Never, to my knowledge, saying Christian or Christianity, doubtless out of concern not to create any new pressure group, Saint Paul announced the disappearance of the ancient human, referred to its groups as well as to their genealogies. So abandoning these formats implied for him abandoning the corresponding belongingnesses. He also quit the laws that had formed them.

Hominization

This new enterprise of universally spreading a subjectivity that's not referred to a culture, not tied to a language at least since the Pentecost, not attached to some genealogy, not obligated by any contract …, I'm not saying that Saint Paul had completely mastered it, nor that he didn't have any predecessor, such as

Socrates, Joel or the Stoics, nor that he had made it become reality immediately in the social and historical concrete, I am merely saying that I read the most powerful appearance of its project in his Epistles. A project that's so originary and so long term that its gesture exceeds its local date and inscription so as to have participated in human destiny ever since its emergence; for the aforementioned 'new creature' bifurcated here from the past the way every branch of the Grand Narrative does. This advent participates, upstream, in the evolutionary time of hominization; downstream, its newness has always remained virgin for two thousand years, still and especially for our time, in which behaviour and discourse always overflow with the archaic libido of belongingness, so powerful, so blind that at the risk of racism, everyone refers to it by the name of identity!

By a cruel paradox, the Epistles spread under the name of those they were addressed to, Ephesians, Philippians, Romans, Thessalonians or Galatians ..., therefore by appellations of belongingness, whereas the letters all implore to no longer take that social, political, sexual or ritual flesh into account, but to exist as individuals. We only quote the author accompanied with the appellations he urged us to abandon. Perhaps we don't tolerate very well having to think a human race that's globalized because uniquely made up of *egos*; no doubt we sorely feel the passion of this becoming-human, the pains of giving birth to hominization.

Worse, many accuse Saint Paul of anti-Semitism; why not call him, as well, anti-Greek, anti-Roman or praise him for anti-philosophy? A Pharisee, a speaker of Greek and a citizen, Saint

Paul took up his origins, often proudly, and emphasized ten times that he remained Jewish and respected the tradition.

He even taught that one should not kill the father and, better still, that one should love him. Should we accuse Einstein of anti-scientism because he amended Newton's laws? He took them up in quitting them; Saint Francis, Luther, Calvin, Lamennais … likewise tried to take up the spirit of Christianity in renewing it. The branch doesn't kill the stem but is supported by it, even if it leaves it. Saint Paul himself invented the floral image of this book: 'If you want to glorify yourself, you do not bear the root, but the root bears you' (Rom. 11.16-18). Writing the Greek language and lodging an appeal based on his right as a Roman citizen, he remained this or that, but on condition, he said, of not glorifying oneself because of it, which means, in his language, not to ceaselessly refer to it in order to define oneself thereby.

The libido of belongingness incited the majority of crimes in history: once erased, peace can occur. Have we ever needed any other message than this one, irenic and liberating? It has to do with inventing a new humanity: humanity, simply.

Event, advent

The Acts and the Epistles say many times that Saint Paul was converted on the road to Damascus. Serious historians aren't fond of the story of Archimedes's bath or the one about Newton's apple. Even so, narratives often speak better than systems. On the dusty route, from the vertical …, the light traced an unexpected crossroads, a sudden branching, one invisible to his companions. The bifurcation transformed the three-formatted Saul.

Prior to this event, whose memory twenty painters, musicians or poets have celebrated, we have the authentic advent, a historical one since Paul himself testified to it, decisive. Seated on the clothes of his accomplices, Saul witnessed the lynching of Stephen. The rocks flew; the victim cried 'I see the heavens opened' and, ripped apart by impacts, died. I won't describe this spectacle in detail, so enticing for the hominids we still are, enjoying spilt blood; whether new or old, none of our formats has yet to erase this ancestor within us. With the distance his observer's site gave him, Saul saw the consequences of the law, downstream, as well as upstream, the construction of belongingness on collective violence. What would Saint Paul soon write? I release you from the Law, he said, that is to say, from the Flesh, that is to say, in part, from social belongingness; free yourself from the Law, from the Flesh, that is to say, from Sin, that is to say, from Death … Resurrect … Quit these burning texts; return to what Saul would no longer see as an event, but as an advent; look, with your eyes wide open, at the bloody act I haven't described – what do you see? Not merely persecutors enjoying the libido of belongingness, but above all the true beginning, yes, the primal scene of every collective and of every subjective. Violence organizes and welds a group together; in the middle, passion and death give rise to an individual subject. Everyone around Stephen: this is the *we*, according to a law whose letter in fact kills; as for the rocks, they conceal the subject, *sub-jectus*, literally thrown underneath. The advent of the *I* under that of the *we*.

Young, witnessing a similar lynching, out of post-war vengeance, I emerged from it newly good.

The three contingencies of universal consciousness

To construct this subject, the Epistles give a new meaning to three terms: a verb, a noun and a subject.

Credo no longer means to believe, in the sense of opinion, confidence or conjecture (πιστεύω, πίστισ) [*pisteúō, pístis*]. I wrote these two words in Greek and Latin in order to better specify that the verb 'to believe' does not translate them. So here is its meaning: supposing that 1 designates objective truth or, on the subjective side, the certainty or conviction this truth brings about; supposing as well that 0, on the contrary, designates objective falsity or the subjective rejection of such an error; then, 'to believe', in its new signification, will now signify travelling one's entire life, hesitating and vibrating, the segment that separates and unites them. Faith ventures into this contingent trembling. 'We walk in Faith, not in clear reason' (2 Cor. 5.7). *Fides* no longer designates good faith or contractual confidence in the word given to or by others, a term from anthropology or Roman law, nor that *bona fides* revered by Latin religion, but a contingency in which are mixed certainty and doubt, conviction and its negation, light and dark, knowledge and ignorance, indeed, this trembling madness unknown to previous ages. Who can doubt like this more than the Son who, at the point of expiring, cried out his dereliction to the Father?

An act that's irreducible to any collective reference, the new faith creates, in return, as it were, the *ego* that becomes its subject. Implied or understood (as though it still lay underneath the rocks), the first word of the Credo, *(ego) credo*, in the end defines the universal subjectivity brought about by this vibration and its perennial swinging. Who am I? The contingency of my faith. I am the one this faith is going to justify, cause to live and save. Again, who am I? The very opposite of assurance; a fear that trembles between being and non-being; in short, a consciousness. The exit from every belongingness. This is how modern consciousness, single, double, multiple, trembling, thrown into time and eternity …, was born. It reverses mastery. As a result, Saint Paul invented, as a writer, the admission of self, shy and without glory, and, before Saint Augustine, the confession of his life, or, at the same time as one or two others, contemporaries of his, the autobiographical novel. This *I* constitutes its existence through what are called the three theological virtues: faith, hope and charity, which precisely describe the three contingent constituents of the new Human. Faith constructs him first.

Hope moves him. No one understood hope better than Charles Péguy, who dressed it as a little girl running under the feet of the grown-ups during a walk and ceaselessly going from one to the other and in this way travelling the path twenty times. You adults are planning to go to some particular place; she comes and goes, moves forward and moves back, joyful, blind to the objectives of your wandering and full of a youthful energy. A driving force, hope impels and sweeps along. Where? Who can say? Does hope, without any doubt, guarantee access to the life triumphant? No, it only promises it, anticipates it certainly, but doesn't ensure it. 'Whoever plows should plow in hope' (1 Cor. 9.10): does he know if he'll harvest? Hope vibrates, like faith, and like it doubts heaven, and strives, timid, to live eternity today. It sculpts time, shapes it and stretches it out.

Faith and hope experience time as saturated with advents, events and beginnings. Faith and hope plunge the *I* into this time of advents. They take it out of every format so as to let it fly towards newnesses. Faith: the *self* becomes established, without any assurance, in contingency. Hope: the *self* moves, without any assurance, and lives, tense, in and from contingency. Thus contingency founds and forms modern consciousness.

In calling these first two constructors of the subject virtues, it would be better, in order to understand them, to move from the ethical and theological sense of this word to the sense of operators suitable for creating the radically new subject. Charity, lastly, fills relations with others with love. Reversing the political or juridical contractual connections of the old belongingnesses, this total bond to others is formed in complete doubt concerning reciprocity: whatever the response it may be given, aggression and spit, insults or rocks, indifference, scorn, enmity, amiability ..., charity always loves, 'pardons all things, *believes* all things, *hopes* all things, endures all things' (1 Cor. 13.7). Its omnitude integrates faith and hope and plunges, with even less assurance, into the contingency, fluctuating and dangerous, proper to relations. Adventurous and generous (of the same family as *genre humain* [human race], the word 'generosity' repeats the term 'gentile', used in the expression that defines Saint Paul as the Apostle to the gentiles, that is so say, to foreigners), this integration allows the new *I* to be in relation with the universality of humans, whatever the origin that may be advertised by their belongingness.

The new *self* is constructed from a triple contingency: faith and doubt; hope that will be happy at an indeterminate time; bonds of unconditional Love. Three weaknesses, three strengths. Less than two millennia after Saint Paul, Descartes attempted to found the subject anew by seeking assurances. I doubt he could have succeeded since contingency and lack of assurance, in short, doubt itself, preside over its birth and its formation. Montaigne described its distraught tremors better. Faith, hope and charity describe the non-ontology of this new subject: its non-settling-into-a-home, its non-assurance, its non-being, its nothingness … the unbelongingness of the soul …

Credo and *cogito*

I don't truly know what I am saying when I say 'I think', but I no longer know at all what I am saying when I say 'I am'. The *cogito* departs from the uncertain so as to conclude with the obscure.

When the first Christians said 'I believe', they suddenly knew, whether slaves or senators, metics, women, Jews, Greeks, sailors or farmers …, that they no longer belonged to this or that class … but rather existed as singular individuals, alone before God and everyone equally by the grace of Jesus Christ. Moving in this way from a category to universal subjectivity, they were resurrected. Burning with charity, they hoped and believed in he who recreated them. The *Credo*, effective, precedes the *cogito*, vague. The latter doesn't conclude as much as the former.

The evaluation of the *I*

Courage alone separates *I am* from *I have value*. 'I don't know who I am' is most often understood as 'I don't matter to anyone.' Quitting the isolation of pure identity, the *I* throws itself into the network of relations of belongingness. Through these, I can evaluate if the fortune of this person surpasses my own; in the arena, the fight decides if the strength of one person wins out over the strength of some other …, but we will always find, elsewhere or tomorrow, some third person who is more powerful, richer, more intelligent and beautiful …; next, cheating often crops up in the counting, the pugilism, the judgement: some prize crowns the one chosen by a pressure group more often than for his value, which is assessed by comparison.

Amid the fluctuating network of these relations, where are we to find the yardstick, the unit – gold, the metre … – objective references of measurement? In brief, does a format exist? Nowhere. The only things to run along these tangled links are comparatives and superlatives, always relative. The only way to evaluate, relation opens up to this relativity; measurement depends on comparison, from which comes all the evil in the world. All value reduces to a size established by society.

The scale collapses. Value, vapour. We fight one another for a shadow.

Thus the absence of format is demonstrated for the *I*. The true measure says and means: zero, nothingness. In the expression 'I am', neither the subject nor the verb signifies anything. To say 'I am who I am', *I = I*, only repeats or says the null relation. I am nothing and am worth nothing. The virtue of humility, the first virtue according this de facto truth, shoots forth from the principle of identity, founding all formal truth. To say 'I am great' is the mistake, one later founding a vice. Satan is. Diabolical ontology. In me, I feel humility to be a first virtue, doubly true. Ethics has its source in the first principles of cognition.

Achilles, stronger than ten soldiers beneath the Trojan walls, Ulysses, more cunning than the Cyclops …, I'm not aware that any of this lot had ever discovered, known or practised the radical solitude of the singular *ego*. The *Nicomachean Ethics* begins by describing a pair of scales: antiquity compared. When the shepherd Gyges discovered, alone, in a deep cave, the ring whose stone rendered him invisible, he profited from the windfall to act without showing himself to kill the king, sleep with the queen and seize the crown; as soon as he became radically alone, he sacrificed to collective values, wealth and hierarchy. Solitary, all the more institutional for being isolated, he immediately jumped into social relation, quitting a possible identity so as to launch himself into belongingness. But, in becoming rich and powerful, did he improve? Plato noticed the opposite instead: the person whose morality is no longer regulated by society becomes the worst of social bastards. Is it better to never stop living in collectivity?

With the *Credo*, the *ego* is born from an inexpressible transcendence, that of the Father, in comparison to which no one nor anything greater can be thought, absolute superlative. From this unit of measurement, we cannot conclude anything except the nothingness of the one who appears to stand its judgement: the relation to infinity reduces the one who compares himself to it to nothing. I can't be the Father. I only exist by grace. Yet, here is the news: from the Father an event of filiation ensues. Coming about like me, his Son resembles me in contingency, unexpectedness, poverty: born on straw, wandering like the homeless, condemned to the lowest of tortures, descending the scale of every collective value all the way down to zero. The incarnation makes the assessment from just now reality: the abstract zero accompanies the lake fishermen in flesh and blood and dies between two thieves. *Ego* equals zero. At first calculated quantitatively, non-ontology thus passes, as true, to real existence.

Let's return to relations. They are no longer used to measure value since this latter remains definitively null; Saint Paul constantly called himself an abortion, rubbish, debris; when I glorify myself, I may only glorify myself in the resurrected Lord. Abandoning all comparison, and therefore assessment and competition, the relation to others, freed, sees another flow run: charity. This latter presupposes the extinguishing of comparison, of the scale by which values are measured, as a necessary but certainly not sufficient condition. 'Patient, helpful, charity does not envy, boast or puff itself up, does not seek its own interest …, pardons all things, believes all things, hopes all things, endures all things' (1 Cor. 13.4-7). Its omnitude includes everything.

As for the null residual value, if God, infinite, gives me his grace, it can become infinite. Freely infinite. In its emptiness, zero becomes capable of receiving the infinite: I am nothing, but I can defeat death. And since I am now no longer afraid of it, humility initiates courage, that is to say, ethics in its entirety. Infinite, ethics is constructed starting from zero. 'We carry this treasure in jars of clay, so that such a great power may be attributed to God and not to us' (2 Cor. 4.7). Made of humus dried with humility, my earthen pot, immanent and contingent, contains a value without scale or format, stemming from transcendence, which demands I take care of this treasure.

Paul, son

Let's return to relations. Who dictates this error-free law, with hundreds of clauses, whose rule formats gestures and the minutes of the day? Who states this exception-free truth attached to thought, to behaviour, to the universe, to the global system of things and humans? Who says this injustice-free jurisdiction and this politics, both applied from the City all the way to the furthest reaches of the inhabited world? Who therefore can obligate the just, the true and the powerful in this way, if not the just, the veridical and the all-powerful: the prophet and God the Father; the wise Father and philosopher;

the judge Emperor and Father? Paul carries a universal trinity of universal fathers on his shoulders. Right before saying 'I am what I am', Paul didn't say 'abortion' for no reason (1 Cor. 15.8). He also repeated 'adoptive son' (Gal. 4.5), not as a rhetorical figure of speech, but in pure truth. For by releasing us from the law, wisdom and jurisdiction, he quit the corresponding fathers and wanted us to free ourselves from them. Contingent, grace and faith replace necessary law; madness and weakness replace wisdom and strength.

Who, consequently, appears gracious and non-legal, mad and unwise, weak and not powerful? The son. Ill-born, after having collaborated in Stephen's execution; born of a Pharisean father and Roman citizen; born again at the feet of Gamaliel; born once more in the middle of the road to Damascus, where he saw the Son. Aborted, adopted, a prodigal son, a traveller, wandering even, he abandoned the powers and veracity of fathers …; yes, I read, dazzled, the Epistles stating, for the first and one of the rarest times in our history, the discourse of a philosopher-son. Before him, prophets, sages, scholars, jurisconsults … played, on the stage of the universal, the role of father; regard with what enthusiasm Plato rushed to the home of the tyrant in Sicily and Diderot to the home of the Czarina Catherine … But also, after them, philosophers and scientists, intellectuals and pontificators …, all of them, over and over again, as fast as they possibly could, took over the place and figure of the father, possibly after having killed their own fathers. Being right, seizing power, judging; conversely, criticizing, destroying, let nothing remain of texts but ashes. Always power, never knowledge.

I have never read anyone but exemplary fathers; I have been educated from childhood by words that were never wrong … I never heard anything but reason and terror.

I have encountered the abortion and adopted one with gratitude. I resemble him at least regarding the weakest points: the son is not always right, doesn't know everything, seeks, stumbles, wanders, makes mistakes, retraces his steps, risks error, blunders, the whip, the rocks under stoning, storms and shipwreck, hunger and thirst, prison, solitude, being let down in a basket along a wall of confinement …, a fragile clay jar pressed from all sides and not crushed; persecuted, abandoned, only knowing how to hope, not in despair; knocked to the ground, but not annihilated … Saint Paul lived as a son, thought as a son, acted as a son, at least three times over, in relation to his three fathers, before whom his failures piled up, persecutions, derisions, tribunals. The son's faith replaces the father's law and truth; the son's hope replaces the father's assured certainty; the son's charity replaces the father's power. But far from killing him, he listens to him and prays to him: 'For the Spirit you have received does not make you slaves so that you will fall back into fear. It makes you adoptive sons and allows you to address God by calling him: *Abba*, Father' (Rom. 8.15).

We live, suffer, think, wander, learn, invent as sons …; here we see the universality of the *ego*-son, which even Descartes wouldn't know, for it plunges into the trembling of faith, hope and love. The philosopher-son haunts the tent of contingency, whose edges flap in the wind. I didn't understand why we lived in the era of the son, I couldn't enter into the theology of the son ... before finally encountering a philosophy that I had never grasped, quite precisely because its author didn't present himself as a father. *Paulus*, weak, little: son.

The adoptive son

Not the family son, but the adoptive son. Genealogy totters: content with his foster father role, Saint Joseph didn't beget; Jesus invoked his and our Father, in the heavens. Father and Son quit their place, their tie, I was going to say their rivalry. We hardly hear the brotherhood of James spoken of, whether metaphorical or carnal; erudite historians will continue to passionately debate it – but what does it matter, really? As far as the mother is concerned, carnally inescapable, but remaining virgin after giving birth, scandalizing many, this innocence partly erases her motherhood. Renewed, the natural scenario of generation turns into adoption, in which dilection's choice replaces blood ties (*Hominescence*, pp. 134–8). For a philosophy, let's say daughter, to come about and think, genealogy must be rethought. This undoing of the ties of blood by adoption, an arrangement legalized by Roman law, favoured the universalization to the human race of the promise made by God to the patriarch Abraham: in order for everyone to participate in chosenness, it couldn't only flow from Sarah's breast.

Conversely, returning to blood ties recently caused us to regress towards archaic illnesses. For modern Western thought has counted its time, from the origin, starting from this adoptive genealogy. Thenceforth we were born neither from the land nor the flesh, but from free will and adoptive dilection. Upon this time-counting is founded a new era, a new consciousness, another cognitive mode: science. In the interminable process of hominization, we stop defining humanity; we adopt it. Truly, we fabricate it.

The Prodigal Son

A long time ago, the son had quit the Father. Maybe this latter had even chased him out, him and his companion, from the first paradise because they had sinned. *Felix culpa*, a happy fault: Eve liberated us from a formatted paradise. She and he, since then, have travelled in a thousand countries in which different languages were spoken and strange rites practised, have learned foreign knowledge and changed skins beneath harsh horizons. Toiling by the sweat of their brow, suffering, childbirths in pain, adaptations, wanderings … They return.

Moved, deeply affected to see everything again without recognizing it all and that they had been more or less forgotten, except by the Father, always there, grown old, attentive, as flustered as they were. Festive joy, jubilation and reconciliation. The erasure of old hatreds, never mentioned, reigns. So much does it condition and signify life, forgetting goes without saying. No more sin, no more law, let's kill the fatted calf. Isaac returns to Abraham's bosom, and both of them sacrifice a ram whose horns were tangled in the stems of a neighbouring bush. Following the lines of the Prodigal Son (Lk. 15.20-32), Eve and Adam return to the paradisiacal farm and to the Father who forgives and presides at the Feast.

Ecclesia

A family feast. Conceived by the first Christians in the first century, did the Church not want to generalize Roman citizenship, already widespread, since legally and politically historical, to the universality of the human race by adding to this citizenship the relations of adoption and love distinctive of this adoptive family, under the eye of God the Father and of Christ, his Son, our brother? Jews plus Latins plus Greeks plus Barbarians from all nations, all men, all women, children and slaves, free men and metics …, new *egos* thus invited, without exclusion, into the set of all the subsets … and entering it easily since they were null and had no properties …, with property always defining a particular subset, that is to say, a belongingness …, all of them, as I was saying, whatever the language they might speak or all of them speaking in tongues … entered then into the chosen people, re-melted under the sign of his promise into the family and love of a Father, both melted into Roman citizenship, law and city, a citizenship itself lastly melted into citizenship of the world, a totality already advocated by the Stoics. Under charity's tongue of fire, Saint Paul liquefied, if I may, the ancient formats in order to let the new branch emerge from them.

Is this universal concept not preparation for what we now need in order to advance towards what we arrogantly call, as though we had invented it, globalization? The disappearance of properties in it erases all the libidos of belongingness. But how are we to found without passion what every passion fights for?

The power of death and the resurrection

So having returned, is the son going to take the father's place? No doubt, for reasons of age, of responsibility, as to others through love of a woman, fatherhood comes to him. By adoptive dilection, he has 'children' in Corinth and Philippi, among the Galatians and the Romans; he feels a paternal love for them that inspires real tears when they wander in their turn. He is a father here. Has he, for all that, quit the place of the son?

No, he never tries to kill the father. Neither Jesus nor Paul, both of them sons, the one in flesh and blood, the other in theory, both of them adoptive in some way, advises parricide, the way Plato subjected Parmenides to it or Oedipus did Laius, the way we believe this act to be inscribed somewhere in our bodies. Each of them teaches to love the father the way the father loves the son. 'Existing in the form of God, Jesus Christ did not consider being equal with God as something to be used to his advantage, but stripped himself of it by taking the form of a servant, by becoming similar to humans, and, having appeared as a simple man, he humbled himself, making himself obedient unto death, even unto death on the cross, which is why God sovereignly raised him up …' (Phil. 2.6-9). Forgiving each other, the son and the father love one another; seated eternally with the one to the right-hand side of the other, they honour and glorify each other mutually. Paul exits the format according to which becoming father means killing one's father and behaving like him afterwards. Before reading him, I hadn't understood, cognitively, how a philosopher-son thought nor what the religion of the Son signified; the entirety of the West descends from him and finds itself in him.

Enslaved to the formats of reason, the master repeats. Regulated by the dialectic, like mechanical dolls, the master and the slave, in an apparent struggle, in reality obey the empire of death, each of them behaving like its slave. The master only rules through death and will only dominate through the terror induced by it. Saint Paul saw the death beneath the law, flushed it out, wanted life and therefore desired never to reign as master. As its divine model, he underwent death and didn't give it. If there is a Lord, here he is a son, like me, like you, like everyone. If a father exists, he is absent from here. In transcendence and eternity. The real world only knows sons. There, abandoning rule, law, format, necessity …, the sons abandon the death implied by them. Therefore they resurrect. How do you become son? By doing away with the law of death. Resurrection, the end of the rule of death. The proof: the Torah and biblical prophetism, Greek *logos* and science, Roman law, lastly, persist and didn't die, like all the rest of antiquity. Paul releases from formats but doesn't destroy them.

The Acts recount that Paul escaped from Damascus by having himself lowered down the city walls in a wicker basket; that he fled safe and sound from many cities of Asia or Europe, condemned, sometimes stoned, often flogged, excluded and driven out; that an earthquake freed him from his prison; that he disembarked in Malta after tempest and shipwreck ...; all of them stories in which the Apostle to the gentiles escaped death. Thus the novel of his life describes, in acts and reduced models, what he professed in words: the Resurrection. His and our lives fight against death. His faith says that this struggle will succeed. 'Death, where is your victory? Where is your sting?' (1 Cor. 13.55). Did the narrative of the Acts stop without warning in order to avoid announcing his martyrdom and his disappearance forever? I believe so. This absence of end fits too well with these repeated announcements of ever new beginnings for the Acts and Paul to finish.

Dead object

Or us. For we gradually became the humans that we are becoming as soon as, thrown in front of us, death became our single object. Whether Neandertal or *sapiens*, we buried our ancestors in places where the house was founded by the tomb and the metropolis by the necropolis; there can be no habitat without penates ... Let's temporarily define humanity as a subject that throws the dead object in front of itself.

Far from fleeing death, we shaped it in *Statues*. Condemned men, repressed murderers, fascinated by death, we turn it into spectacle and story; it incessantly haunts our representations; maybe there is no representation except of it. Priests, we preside over funeral ceremonies; warriors, conquerors, defenders, we kill; farmers, we bury grain; shepherds, we raise lambs to sacrifice them. Prayers, conflicts, food, our practices concern death. Our knowledge as well: would any knowledge have ever taken place without us holding in front of ourselves, prisoners of our hands, an immobilized object, the skeleton of a schema, a ghostly concept …, without us passing its shade and shadow down from generation to generation? Corpses follow us in such a way that at every moment they haunt, in the cave of our rooms, our screens.

Object: death, tomb, statue, idea, ghost.

The renascent subject

The subject: it is split into mortal, thrown under, like Stephen, and immortal, resurrected. Death freezes our formats, schematizes them, purifies them …; immortality launches our branches. We live in the grip of mortal necessity. Individuals, cultures, humanity, there is no known exception to its rule. We have always known that every human is mortal. I will die. Who am I? This proscribed man, thrown under its law. Condemned, I am only concerned with pardon. We feel and experience ourselves as eternal. We survive as double; strangled by death, striving to breathe, to free ourselves from its herring barrel, having no rest until we dominate it. At least and at the outset, by defying it, with daring bravery, with insolent audacity, with laughter, heroism and self-abnegation. It alone is our object, our emotion and concern, our enemy, partner and adversary; we set ourselves immortality as our goal.

Every human invention has always had this stubborn project as its motor. Every courageous heart, every gathering of plants, every hunting of animals, every travelling, relation, whether one of love or war, celebration, poem, theorem ... is undertaken to survive, either as a person or as a group; every structure we raise, cairn, tumulus, hovel, wall, city and port, refuge from death[2] ... will remain as marks after us since, all around us, only these traces, manuscripts or graffiti remain human. In the tomb, the corpse; on the stele, the immortal. Every engraved word presupposes or projects immortality. We count every transmission (passing on tradition or memory, practice or theory ...) as a piece of this victory. Programmed, plants and animals strike the colours; we hoist them up. They obey, we rebel. Even if partial, this triumph contributes to humanity's adoption, to its continuous fabrication. Every duty of remembrance is supported by nothing but a project of immortality.

Who am I? A subject under the rocks and a project of resurrection, in front of the object, a dead statue and a trace of immortality. Ashes and works; death pangs and recovery; obedience and revolt; ignoble pride and noble humility; *Requiem* and hope ...: a branch with a soon inert trunk and with ever-green twigs. Self-consciousness is woven from a lethal weft and an immortal warp, from a stiff oldness followed by an ever-virgin youth; chance and necessity, memory and remembrance, knowledge and ignorance. My knowledge itself is split between format and invention.

Who am I? This bifurcation. The pairle of the escutcheon, a figure crossing in chiasm. Exhausted, indefatigable. Spurned, passionately in love. Unconscious, keenly penetrating. Earth and air, crawling, flying. Water and fire, incandescent frost. Enthusiastic-indifferent. An athlete and an abortion, living, full of resources. Weight and grace. Headstrong, scatterbrained. Dormant-inert-carnal and awake-nascent-carnal. Lying, standing; pathetic, enterprising. Emotion and abstraction. Me and always other. Mixed-race, completed lefty, Educated-Third. Hermaphrodite. Angel and animal. Statue and music. On balance, crying with joy.

This is the escutcheon of our practices and our knowledge, which are regulated by necessity, defined by impossibilities, worried about contingency, launched into open possibilities, ramified in the square of modality: from the middle of the first diagonal, uniting the two sides of the necessary and the impossible, the second half-diagonal shoots out towards the point where the other two sides, the possible and the contingent, meet. In neither the subject nor its actions, whether practical actions or theoretical ones, can formats and inventions be separated.

The project of immortality

En route. We only have one project, one future, one hope: the inventive victory over necessary and formatted death. Philosophy's project is to succeed in this combat. We have dreamt of this from our origins, ever since we learned how to produce children, dance, sing, talk, make fire and cook, domesticate species that reproduce, make cheese and jams, criss-cross batter boards over foundations, compose music, write, count, measure. Gilgamesh rose up to defy his end; Ulysses and Orpheus, a few returning ghosts, shining with glory, crossed the Hellenic Underworld's river of forgetfulness in the other direction; Thales abstracted geometry from Egyptian tombs, towards the eternity of forms; Jesus Christ resurrected and will return among us … Our entire past history, dreams, beliefs, gestures and actions, faith, hope and knowledge … worked at this. Through passing on, technology, science …, we have already – influenced by religions – taken several advance bastions of its entrenchments. It took us millennia before we read in texts, nonetheless reread a thousand times and venerated, the 'you shall not kill' in its evident, concrete, merciful sense. Woken to the evidence, some of us want to abolish this penalty before the courts, in everyday life and gradually in relations between groups, in which the libido of belongingness still causes as much male rage as in rats; some of us try to extinguish nuclear weapons and seek to pacify wars between fathers. Upon reaching an adult age, sons pardon their father and, through a new revenge, no longer want to kill their sons. So who am I? Son and father. Taking the branch of the road that is reconciliation. A new victory over this last death.

Our species became, is becoming and will become such for having mastered, more than space and things, time. In concrete technologies as well as in the unfolding of the Grand Narrative. At the furthest extremity of this long trip, here and today, the crossroads between death and immortality opens again. On the one side, our own works, with global risks, doubled with competition and comparison, with wealth and poverty ..., on the other and once again, repetition and the newness Saint Paul called Resurrection. Like him, I'm looking for a rupture without death.

Even if this death of death, already prophesied by Hosea (13.14), occurs like a thief in the night, my hearing, sharper than it, hears it coming with much noise ... on the day we understand that there is no death except the one that's organized, desired, decided, celebrated, repeated by formats like those of the recent fundamentalist convulsions that are killing each other ..., on the day when hope for life, through our works, breaks out ...

On the day after that day, we will have to learn how to inhabit the new world projected by our labours and this book.

CHARLES PÉGUY, *BRUNETIÈRE,* FALL 1906

When in a tree, generally in a plant, whether bush or arborescent, for some reason – frozen, a freeze, a gust, trauma, drought – a bud aborts, a growth fails, a secondary treetop or the main one withers, arborescent nature all the same doesn't desperately strive to make life come out of death, sap out of dryness and sterility, riches, abundance, affluence – and poverty – out of poverty and destitution and, as the ancients once said, the humid out of the dry; rather this nature abandons the dying treetop to its sterile fate; it performs a subsumption, an intussusception, an absumption, a reprise; it takes up again more profoundly; a new bud is born, under the first one, often very far under, as far under the first one as is necessary to reach the sources of sap that have remained alive; a new bud, lower, a new bud silently pierces the hard bark, a bud come from the interior and the profound, from the enduring insides of the tree. A secret emissary.

From the Nymphs that lived under the hard bark,[3]

a new branch is being born at the axil of the abandoned branch, a new treetop is being prepared.

In this way and only in this way do trees restore themselves and continue.

The treetop that will wither can still be entirely leafy, entirely proud with leaves like a plume. It has nonetheless been marked by disease. If it has been so marked, if it is infected, it has nonetheless been condemned. And yet it is still full of leaves. But these leaves will wither. In comparison with this treetop, this poor little bud, this little bit of red nose peeping out, which underneath is piercing the hard bark, doesn't look like much of anything. Yet it is this bud that has become the representative, that has been set up as the sole representative of Great Pan in this business. It is from this that salvation will come, that survival and rebirth will come forth.

Particularly, it is for reasons of the same type that, when one takes a cutting, one must indeed keep oneself from keeping all this bushy verdure, all this pile of foliage that seems to form the luxuriance and strength of the plant, that doubtless formed the luxuriance and strength of the old plant, that wouldn't form the strength, that wouldn't form anything but the withering of the new plant, for a cutting is an artificial replacement of the treetop. On the contrary, all of that must be pruned off, and the floor must be given to these quite little buds that are quite ready, in addition: to these anonymous possible buds that are going to commence.

One can see what consequences would result from these observations – we will perhaps examine them one day – in the order of morality, of the social, of work, of productivity, of appropriation, of every culture and generally of all humanity.

What is abandoned is abandoned. No longer comes back. Let's not talk about it anymore. Let's leave that to someone else.

An abandoned treetop is abandoned. Eternally. No regret, no remorse, no emotion. Nature is perfectly unaware of any type of consideration of that order.

Such behaviour is so truly nature's behaviour that it can be confirmed everywhere and down to the smallest detail: for example – and to confine ourselves to this particular case – it can particularly be confirmed – in art, in letters, in philosophy – in the formation, in the birth, in the growth, in the development, in the culmination, in the decline, in the mode of succession of what we have called genres. Far from genres evolving, as has been said by a hasty, cursory, brutal and maybe even a little crude application of a modern metaphysics claimed to be naturalistic – itself forced from a modest natural naturalistic hypothesis that didn't contain it – on to this history of genres, far from having to speak about an evolution of genres, *genres, like all the humanities, arboresce, and one never has to speak about anything but an arborescence of* genres.[4] *When a genre has been achieved, like a humanity, when it has been exhausted, when it has been crowned, no evolution, no transformation, no deformation, no reformation will derive anything from it ever again. No miracle – for this would still be a miracle, clearly a miracle, the perpetual shameful modern miracle – no miracle will make anything come out of it ever again. The place belongs to another, the old place in the sun. Another will come out, another will be born, grow, another will live, a genre bud, a genre budding, starting small or coming out in an eruption, will come out freshly, will boldly and directly make its way, its truth, its life. Arborescent nature is not the art of making use of the leftovers. The new branch, the new treetop, the new genre is not from the old branch, the old treetop, the old kneaded, filtered, triturated, manipulated genre. Redone, reprised, corrected, revised, augmented, diminished. No, it's new. Quite simply a new genre. It's a new thing.*

It no longer has, it doesn't have in it the old elements, tinkered-with, of the old genre. Those who do such tinkerings are highly gifted men, inordinately intelligent, clever devils, people who have enormous talent. Genius doesn't proceed in this way. Genius doesn't proceed by such tinkerings.

It doesn't proceed by kneadings, decantations, triturations or detritus. By macerations or preserves.

Genius proceeds much more simply. Or rather it proceeds absolutely simply. It proceeds with an absolute, total, infinite simplicity. In no way by transmutations, cookings or residues – or the selling and buying of the inedible offal coming from food supply services – but by constant renewals, refreshings, reinventions, reintuitions, re-sourcings. Genius being of the order of nature, genius's work is of the order of nature's work; every elaboration of genius is a natural elaboration; every invention, every renewal of genius comes by means of this arborescence we have recognized …

Œuvres en prose complètes, tome II, Bibliothèque de la Pléiade, Gallimard, 1988; pp. 583–5.

Narrative

Event

Ordinary peoples, criminal and peaceful, had farmed the low-lying plains to the north of the Dardanelles and the Sea of Marmara for generations. Suddenly, the Bosporus cork gave way. A mass burst that no one had realized formed a dam, and, stemming from the Mediterranean, brackish torrents poured out in cataracts into this low country, inundating furrows and villages as they passed, likewise killing humans and animals. In less than 'forty days and forty nights', a small lake widened to the dimensions of the Black Sea. This event took place in the now dated times of deglaciation, during which similar populations, residing at the bottom of the Strait of Dover, witnessed, just as powerless, an irresistible invasion of water. Whether geologists or prehistorians, some people imagine that the Bible recorded the Eastern European event under the name and via the story of Noah. For over the course of the preceding weeks, an astute sage could have heard the high barrier cracking and persuaded his family and friends to prepare for catastrophe; Noah, a son saved amid the dead and a father of the survivors.

How should we define an event? As that bomb whose contingent newness interrupts a state of affairs that has been formatted for a long enough time to make people believe in its perenniality: a people was going about its business in calm, it disappeared; one family alone remained; an ancient lake of reduced dimensions expanded; a fragment of history bifurcated.

Rare, news of this type astounds when it is told; it tears apart the old formats.

The consequences of the event

Supposing the experts were not mistaken about the reality or the date of this event, nor about its interpretation, not only is the rupture of the Bosporus interesting to those men or women who were unaware of it and thereby learn that such climatic fluctuations can return, but this rupture in truth also had an enormous impact: it destroyed the civilization whose ruins we discover on submerged shores; it reshaped a notable part of the global map and opened Russia up to the sea by giving rise to new maritime exchanges; it staged a hero whose role as a new Adam caused exegetes to ponder …; in short, it renewed our view of time and the face of the Earth. Strictly physical, the event had a historical and religious effect.

When, translated into the majority of languages, the Bible penetrated cultures and mentalities, the story of Noah (the deluge, the bank-ark of visible living things and the invention of the first biotechnology, wine, through domestication of an invisible ferment …) spread even farther than the waters. Supposing, I repeat, that the aforementioned rupture truly took place, there and during the times stated, the consequences of the event surpassed the simple mechanical effect, whether physical or one of terror, that could be brought about by cascade or cataract. The verb 'surpass', used intentionally here, doesn't merely designate the quantity or volume of action, but also a change of nature and quality: the appearance of space and the direction of history. The map of Eurasia was redrawn; some people even started human history from zero or saw it deviate. Global and cultural consequences ensue from a physical and local event: from one format, another one.

Causes

Are there other events, comparable, that, by their strength, seem to be an exception to the usual chain of causes, in which consequences are equivalent to their conditions, in which the entire effect is found in the full cause? The physics of the Earth dates five eradications of the majority of living things fairly exactly, catastrophes due, it seems, to volcanism or an aerolite strike, a normal eruption or a chance impact bringing about what could be called a nuclear winter; the dust made to shoot up into the atmosphere by these occasions, put into orbit, plunged the globe into a long and icy night. Local cause, universal effect; physical cause, biological effect. The consequences bifurcate in nature as well as in scope.

What should we call an event? When known causes unfold in such a way that the expected effects remain similar to what precedes, the sequence plunges into a format that's predictable by the ordinary rule of causality: the hours follow one another, duration flows; everyone gets bored or lives their share of happiness. But should a colossal occurrence suddenly arise that brings about unexpected effects in size or nature, and should the monotonous format of previous rules deviate, in direction for example, then we call them events.

I am now talking about them and newness.

On little causes

Am I mistaken? The millions of horsepower unfolded by the Bosporus rupture, the impact of an aerolite in Mexico or Siberia, ten volcanic eruptions in Iceland or the Sunda Islands can, by their power, produce a thousand devastating effects, and universal to boot. So these events could be used as examples of the normal development of the ordinary format: tremendous causes, enormous effect. But there are other ones where the strength of the causes diminishes down to the minimum and even to zero and whose effects nonetheless surpass all proportion.

Thus history, both Roman and universal, deviated at the sight of Cleopatra's nose, whose curve was appreciated by Caesar and Antony in turn.[1] Causes with almost zero strength or actions that are ridiculous today, becoming decisive tomorrow, will produce effects that shake singular existences and global empires. Who can be certain that the Soviet regime wasn't shaken at all by the Vatican institution, so weak that Stalin derisively asked how many divisions it had? Who knows how to weigh the power of a symbol that is apparently without any strength?

Can we assess the consequences of words? A piece of gossip grows into rumour and, spreading into slander, drives a victim to suicide. Who can predict this propagation? The remark kills: unexpected, the effect surpasses tremendously an imponderable cause. Do words change the human adventure? The fact that reading shapes the body, decides actions and enchants the world is shown by *Don Quixote*. Historians have trouble filling this gap between heavy consequences and the lightness of language. Who can evaluate the effectiveness of an announcement, of a lie, of a truth? We rarely master the effects of our productions, whether words or things.

Nature and cultures

We also don't always know how to assess the consequences of our equations: words and phrases from another language. Newton discovered universal gravitation; ever since then, no one sees the sky or the stars in the same way, nor the Earth or the apples in the yard. Conceive of grace as the opposite of gravity. Three letters, a sign and a figure ($e = mc^2$) provided us with the access key to forces from which we drew bombs destructive enough for the terror caused by them to change international relations. In manipulating the atom, the chain of genes or cloning bacteria ..., aren't we playing with fire? We have risked fires ever since the supposed domestication of flames; neglected, a match can set hectares aflame.

We shall soon laugh at those who told of mastering technologies. Who can guess if and how some object, common however and come from our industry, can be diverted someday into a symbol, an icon, nay, into a divinity? Our ancestors venerated, it is said, fetishes-logs that they had just cut; worse, they sometimes sacrificed their children to them; the blind worship of certain products of our economy kills our families on the freeways. Fearing seeing their statues move, the Greek sometimes covered them with chains. What could be more commonplace than to adore idols sculpted by our hands or banal stars on brightly coloured advertisements?

A flint knife hunts but also murders. The pursuit of aurochs helped them survive; was this pursuit running towards the eradication of the species? Who knows? What we shape and think we master departs to seek its fortune in the world, being born to a life of its own. The anxiety attached to the story of the sorcerer's apprentice has affected *Homo faber* from its first productions and haunted our technologies and our sciences all the way up to this morning.

Generalization

The reader might be surprised that the preceding pages went from casual slander to scientific formulas, and lastly to our fabrications in general. All these examples are united by the break in proportion between little causes and gigantic consequences, whether favourable or disastrous, a union resulting from linking the scale of information, so delicate it plunges down to the minuscule weaknesses that are verbal utterances or psychological energies, to the scale we started from, that of earthquakes.

For we also don't always know how to assess the consequences of a purely physical phenomenon: thus, before chaos theory, Poincaré demonstrated that the Earth could, without warning, leave the solar system and depart, it too, to seek its fortune in the world. So the cause quits the enormous so as to descend, even in mechanics, to the imperceptible and join there, in the minuscule, Cleopatra and slander. Laugh at historians who remain deterministic in human affairs, while the hardest sciences accept that unpredictable effects linked to initial conditions that can't be perceived by the most minute observation can occur.

The loop closes: the same disproportion can affect inert and human-caused phenomena just as much; from the most imperceptible to the astronomical, from nature to cultures, on every scale of force ..., the possibility of a gap between cause and consequence can be found. A tiny mutation can lead to the emergence of a living species whose numbers will occupy the globe. The concept of event becomes universal. While it used to seem so slight and circumstantial ... that, to express these qualities, we said 'event-oriented' [*événementiel*], it is now losing its character of being an exception so as to join, if not a rule, at least a crowd. This book celebrates the access of contingent singularities to the universal. Narrative unites with the law.

The observer and his interest

I just talked about things themselves as though no one was perceiving them. Yet an event is measured in relation to the interest an observer takes in it. If he gets bored, he will run to the news; so an unexpected announcement interests him: yesterday morning, Santorini exploded, destroying the Minoan culture; yesterday evening, Newton invented universal gravitation ... The interest increases with the newness; the subject no longer gets bored.

But how does he recognize an event? If the occurrence that happened maintains no relation with his previous experience, will it interest him? Who among his family gave credence to Noah's apocalyptic opinions, alone with his ark on his hands? There was no prophet in the region ... But shepherds must also have already, for a long time, raised livestock on the narrow shores of the old Black Sea for the cataract to have devastated their history. Please reread, above, the two first sentences: ... *for generations. Suddenly* ... Here are the two acts of the event: before it, a kind of monotonous format reigned, inducing habit or boredom, and, all of a sudden, a contingent break occurs in this regime. Is it truly a matter of a totally new thing?

Not really: every bifurcation displays two stems. Scientists before Newton must have already asked a few questions and missed the right answer regarding the cause of motion; houses, palaces, a noteworthy social organization and a few fumaroles must have appeared on the island of Thera; the victims of a love-at-first-sight thunderbolt must have experienced two or three things regarding love … Does newness descend from the sky, indecipherable? Not really. Granted, the Greek language was ignorant of the word volcano; granted, astrobiology claims that the first coded RNAs arrived on Earth on board aerolites …, but, most of the time, the news maintains some relation with a usage that precedes it and which the news shakes up. Something exists beforehand, which the bifurcation disfigures. A father dictates the law; his son disobeys it.

When Saint Paul announced the Good News, he grafted the Christian branch onto the Jewish tree and its Pharisean bough; the graft was born on the rootstock. 'If you, being cut off a wild olive tree to which you belong by nature, have been, contrary to your nature, grafted onto a cultivated olive tree, how much better would these natural branches be grafted onto their own olive tree to which they belong by nature!' (Rom. 11.24). In the shape of a ramification, each of these examples, chosen deliberately for their difference in nature or cultures, presents a stem, stable, and a branch, new.

Action and thought

Just as the event, universal and singular, in fact traverses every scale of power, from the colossal to the minimal, so it crosses the border separating the raw occurrence, in which a new thing separates off from the format, separating the raw occurrence, as I was saying, from the observer wakened by the news. He who gets bored experiences the uniformity of a sequence of repeated occurrences; strong and keen, on the contrary, interest suddenly springs up when the newness of the event shoots up from the rule …, when the graft diverges from the rootstock. Leaping from his torpor to the news, the observer finds himself torn and divided on this double schema, along with the things around him, things subjected to a suddenly broken law. The explosion of a seventh chord breaks up the greyness, and the ear wakes up. This book celebrates the waking at the fork-point between the stem and the branch because these days we are living on this double point of tangency, from which the word contingency was derived.

So not only does the observer get up from bed but so does the possible actor. Who can assess the effectiveness of his practices on things, living beings, humans and circumstances? Vast investments prove to be powerless; a flick of a finger decides a triumph. However powerful he may see himself, the master has less power than he thinks; however weak we may see ourselves, each of us has more power than we think. However feeble my weakness may be experienced to be, more strength comes from my arm than from a butterfly's wing; if this wing can trigger a hurricane, what will my fingers be able to achieve? Neither the tyrant nor the slave assesses their scope. Tomorrow, the former will fall, of himself, and the latter will scorn taking power in order to try to establish a less stupid world.

Jubilation, gambling

This ignorance of the effect … inspires hope for action, joyous decision, freedom of destiny. Through the inexpertise it grants me, contingency gives rise to an inexhaustible jubilation of willing, thinking, undertaking. A solitary essayist throws his works out like dice; it's tough luck if he doesn't write *Phèdre*, at least he will have lived, that is to say, attempted. Within every newborn, the adventure of the Messiah, hoped for or already come, is at stake. We don't so much live plunged in a fated sequence of predictable causes, in an incontestably necessary real … as in an extraordinary game of chance in which the actual and the probable can lose their weight of seriousness in relation to the inactual, the symbolic, the unexpected, the invented at leisure, the mad and the weak. Something that, here and now, appears dramatic and pressing disappears into smoke a bit later, and something whose importance no one sees becomes, patiently or lightning fast, the most important thing. Today's necessary quickly turns into the impossible, and the contingent suddenly changes into necessity, the reasonable into imbecility and the insane into rational.

The thinker wagers. The man of action gambles. The artist risks. Cook launched into the Coral Sea without knowledge; Gallois discovered groups before dying in a duel at dawn. When more conventional people engage in what we call actuality, they sometimes lose their status of being a thinker and the effectiveness of their actions. Realism proposes bad bets. If you want to lose your soul, work to save it; for he saves it in the end who appears to have lost it. Play for high stakes. Imagine, invent, plan, something of it will always remain. Fortune, sometimes, smiles on daringness, alone reasonable; learned ignorance thinks.

Four new things

Conversely, are you getting bored? I will therefore announce four events to you.

Here, first of all, is our news: millions of years ago, *Homo sapiens* appeared; quadrumanous, it invented walking; made fire; left Africa; disembarked in Australia, then in Alaska, by the Aleutian Islands; carved and polished stones; a mammoth hunter, boldly crossed the Atlantic, from ice floes to fragments of pack ice, from southwest France to America; domesticated the dog; raised sheep and cattle; farmed corn and wheat; crossbred pigeons as well as apple trees; massively stoning the corpses of kings, built multiple pyramids; prohibited human sacrifice; discovered universal gravitation and non-commutative geometry; wrote the *Divine Comedy, Don Quixote*, the *Essays* ...; composed *Le Tombeau de Couperin* ...; will it practice the only motto compatible with its survival: love one another? ... How should we define humanity? By this narrative of new and contingent events that are unpredictable before they occur but formatted as a semi-necessary chain when drawn descending towards us.

Before the human adventure, the living species, gasping for breath, succeeded one another; almost with the birth of the Earth, a coded RNA arose, maybe from elsewhere, capable of duplicating itself; bacteria reigned for three billion years; multicellular organisms exploded; Burgess Shale dates the arrival of hard parts, already evoked; so the immense tree of kingdoms, orders, genuses and families unfolded, as surprisingly as our own inventions, its multiple ramifications and twigs. How should we define life? By this narrative of new and contingent events that are unpredictable before they occur but formatted as a semi-necessary chain when drawn descending towards us.

Are you getting bored? Only new announcements interest you? But what events should we call interesting? Here you are: adventures, discoveries, life, even time itself … incessantly flit from one thing to another.

Before these two arborescences, let's not omit, for the sake of still ridding ourselves of boredom, the news of the world. Over the course of its expansion, the incandescent universe cooled; having reached some given temperature, ionization stopped, which was preventing particles from fastening to nuclei; matter could then become separated from light; the latter continues, the former concentrates: after the brilliant homogeneity of the state of youth, in which even atoms hadn't yet formed, the distribution of galaxies followed, separated by a semi-void. We know how to date this knot of change, especially ever since we have been able to observe the cosmic microwave background and measure its residual heat. The old stem, present, continues to vibrate; the new branch forms the observable universe. Just as the classification of living things summarizes a temporal evolution in which branches continually surge up, always new, the classification Mendeleev drew up for the simple chemical bodies likewise summarizes the arborescence of their formation according to time. How should we define some metal? By the date of its birth, of its newness. What should we say, as well, about some star, dwarf or supergiant? Red old age; blue youth.

I shall stop, so as not to bore you, this series of examples, nevertheless so interesting; these four new things – material bodies, universe, life, humanity … – can be summed up in a single word, nature. How should we define it? By its original meaning: that which was born, that which is born, that which will be born; that is, a narrative of newborn and contingent events that are unpredictable before they occur but formatted as a semi-necessary chain when drawn descending towards us.

Everything we learn shoots forth; everything we produce surges up; everything that exists invents … The Grand Narrative resembles a universal arborescence of contingent events and of new things. Necessity, where does your victory fall to pieces? You no longer wound with your sting. Has boredom just breathed its final breath in order to leave a cradle for the joyous interest of the new?

Boredom

Why are you bored? Because long chains of reason always go over the same formats again; rocks fall, rivers flow, hours strike, predatory species reign by programme: lion, Alexander; jackal, Louis XIV; hyena, Stalin; vulture, Pinochet …; eagles: Napoleon won and lost battles, England conquered a thousand colonies, America is mastering the world …; each line falls into line with monotonous rules, always modelled on the same format of power and death. Things that are predictable before they occur and forming necessary sequences, without any information.

Such formats, methodical, dominate. Force kills but doesn't invent much. Old, Satan criticizes, God creates the new.

Boredom repeats laws. Heavy bodies attract each other; fire cools; order runs towards disorder; organisms deteriorate; genetic automatons repeat their programmed behaviour. Nothing learns anything. Don't slap, the mosquitoes will come back to harass you; don't harass these humans, they will return to battle. The universe is running to the big crunch, a terminal crushing, symmetrical to the big bang; the Sun is running towards its nova, the final explosion; stem cells are running towards a series of inhibitions; species are running towards their extinction; invention towards repetition; new thought towards litany; works towards commentary; violence towards violence, power towards power, both towards death. Boredom kills.

The deadly melancholy of 'news'

Are you still bored? Run to the news. So what do you call by this common name? Words and images of power and glory play musical chairs in which the name of he or she being talked about changes, but the law for gestures is repeated: conquering power, abandoning it, being expelled from it. To seize it, you must kill. Killing ever since eagles and lambs, Cain and Abel, Alexander and Stalin …, hunting, the struggle for life.

Run to the news: TV channels, said to be informational, at all hours drum out with a monotonous sound the norm of death, seasoned with terror and pity. Murders, corpses in large numbers. Their dismal links in a chain imprison with death. How would news emerge from these melancholic programmes, which are as necessary as the laws of the deadly falling of bodies? True news announces adventure, life, inventions, contingency.

New praise of format

A norm nevertheless presents a semi-interest. A tic, repetition imprisons, of course, but, an adjuvant for work, habit lightens effort. To flee these chains or not, that is a real question.

For if the law binds, it aids, as I have said. If uniformity puts to sleep, if monotony stiffens and subjugates to its drug, they nonetheless regularize. Having two aspects, a clock repeats and announces, rocks us to sleep and wakes us up, makes time pass and sounds reveille, in the day's schedule as well as in biology. Dense with twenty watches, our body suffers from their disturbance when it crosses meridians. Training allows us to progress following the cadenced return of gestures and exercises. Would we learn without the imitation done by mirror neurons? From the body to the work, the same programme: rhythms have the double role of benumbing us into sleep and carrying out tasks. A work requires following a repeated shift for writing. *Nulla dies sine linea*: not a single day without a line. A rule has two functions: enslaving and freeing. Art is born from constraints and dies from freedom. But when, conversely, academicism kills it, it will be reborn tomorrow from untying its attachments. Jumbles of commentary strangle intelligence, which is rendered stupid by absence of tradition. There can be no style without grammar, whose rules don't create style. Wise, measure counts and orders; meticulous, it rigidifies the inventive impetus. Without form, there can be no new work, which occurs outside form.

Natural, clocks follow the circuit of the planets but mark with their strikes heartbeats, coups d'état and coups de théâtre;[2] there can be no life without the law-abiding return of the dawn nor can there be spring without the eternal return to the vernal point, no poetry without rhythm, no harmony without the anharmony of sevenths or style without the breaking of style, no solar system without the chaos that can make a planet bifurcate from its ellipse. Time runs towards entropy or, bifurcating, cadences the expansion of the universe and vital evolution.

Peter and Paul exit prison; the more Jewish one, sensitive to signs, freed by the Angel; the more Greek one, sensitive to reason, by an earthquake. Exiting, untying bonds …, suddenly their News emerges. Socrates refuses to let anyone free him and quits life rather than the bars. Should we flee or seek the law, and which one? Nature is characterized by format and news, death and newness, just as much as our cultures are.

Freedom from death

And once again, what is there that's interesting? That which shoots forth from out of prison, from out of rules and the uniform: the exit. The arrival of a poison, oxygen, killed, but let burning organisms be born. On the backs of birds, feathers launched crawling reptiles high. The Fertile Crescent's monotheism thundered as new amid the idolatries. Abraham, the father, no longer killed the son. Anaximander imagined the indefinite at the origins of geometry. Sometimes, rarely, peace arrives. What is there that's interesting? Arrival. Birth. The rare heat, not the law-abiding cold. Wakefulness; neither sleep nor dream. The invention of survival. The exiting from the cave, the exiting from the tomb. Always think about exiting. About being born to defeat death.

And what else? The information known as news keeps trotting out lethal invariances, the inert laws of gravity or those, animal, of the jungle. But in the precise, yet opposite, sense of this word, information nullifies these repetitions in order to launch rarity. What is there that's interesting? Divergence, emergence. An exception to nothingness: the big bang; to entropy: organization; to the bacterial kingdom: multicellular organisms; to quadrupedalism: erect posture; to the dirty, murderous hands of doctors and nurses: Semmelweis, then Pasteur; to the sinister rules of violence: extremely rare love; to platitude: the work. What is there that's interesting? The exit from death pangs: life, inventive thought, heat, love, benevolent courage. Birth, the victory of contingent life over necessary death.

The theory of four truths

Are you still bored? I will lastly tell you your four truths. For, quadruple, formats and newnesses concern, as I just said, inert nature, life, humans and their productions. Bifurcating from an incandescent and homogeneous universe, galaxies, new, scattered off one after the other; leaving prokaryotes, eukaryotes, new carriers of nuclei, announced a different reproduction for the living things to come; soon upright, Lucy's family to come would leave Africa and travel the world ... Italian music abandoned the Baroque; Poincaré amended Laplace in such a way that the world revolved differently; when will we change our constitution? Sciences, philosophies, arts, religions, politics ... invent newnesses the way genes mutate or particles bifurcate.

Talking in this way scandalizes. How are we to say, with one voice, every type of event? How are we to unite the examples given without distinguishing them? Deprived of the discipline that would synthesize them, this book, an orphan son, lacks the language in which it could express their concordance, that of the newness whose ramification is found in every place and concerns us as well.

Us, in this story

That was the narrative relating events that are material, living, historical or cultural ..., and here now is the very person who recounts, whether me or others. His sentences don't merely objectively follow a succession of states of affairs but can transform the one who announces it and those who will hear it. Evolution changes the relations between particles, the species of living things and the customs of groups; its narrative can change the one who tells it and those who hear it. Does a told event create us?

Certain narratives harbour this hardly described strange force of transformation. La Fontaine evaluated it in 'The Power of Fables' (Book VIII, Fable 4), in which the fable it recounts captures attention. The scene takes place in ancient Athens, when Philip, at the gates of the city, was putting the country in danger. From the height of the tribune, in powerful rhetorical figures of speech, the orator Demades was describing an emotional situation … 'Vain and fickle', the people were chattering away. Consequently, changing methods: "'Ceres", he began, "was travelling one day with the Eel and the Swallow. A river stopped them, and the Eel, by swimming, as well as the Swallow, by flying, soon crossed it." The assembly instantly shouted as one, "And Ceres, what did she do?" – "What did she do? First, a swift wrath angered her with you."' Filled with ire, Demades then drew his fellow citizens out of apathy.

How did the orator transfix the assembly? How did he strike up attention? Narrative succeeds where warning, the violence of words, eloquence, peril all fail. Narrative, yes, but which one? Recall the phrase I began with, already quoted: … *for generations. Suddenly* … In La Fontaine's super-Fable (I am calling it this because it evaluates the power of fables), the same rhetorical figure appears: … *was travelling … A river stopped them …* They were going about their business; the Flood killed them. They were walking along a path; the river blocked them. At a point, a discontinuity breaks a continuous path. There, a catastrophe of cataracts, in a few days, interrupted centuries of ploughing;

here, a river crossing the path stopped, at a point, the trip. A dam breaking or another blocking, to be crossed, tears a system of transit apart. A newness blocks the law. The event cuts off the course of the format. This is the knot where the branch leaves its trace on the ordinary trunk. Interest grows with the intersection that forbids: yes, read the ramified form of the narrative three times: in what it recounts; in the form by which it is related; in the soul of the person who listens to it. I flit from one thing to another, therefore I listen.

Better yet, the fabulist appeared in person; he said: 'If "Donkey Skin" had been told to me, I would have taken an extreme pleasure in it.' What pleasure, ye gods? Must the new event also change me, the teller, orator or rhymester, who changed readers and listeners so well? The Athenians woke up, became moved; the narrative recreated them …; I never cease to delight in 'Donkey Skin'; even if I know it by heart, it recreates me. Our soul …, does it resemble a tree exploding with branches and bouquets? But what pleasure, oh ye gods? That of attention, which metamorphoses beasts into beauties. This new state appears at the point where the branch is grafted onto the stem. This graft transforms the law repeated in the stem into another course; sighing with boredom, I shake myself; sleepy, I wake up from my dogma; imitating, I invent; a trunk, I become a branch; fastened into the donkey skin, I exit. I convert. Dead, I am reborn.

This newness rectifies the course of internal time, whose flow murmurs with vocalises and phrases. This course bends with the narrative. Suddenly the sense changes sense; sense is born when sense changes, whether it is a question of direction or signification.[3] When a flow bifurcates. When a donkey becomes a girl. What song, what spell, what metamorphosis … make these mute animals talk in the *Fables*? Silent, do children also learn language there?

Old, I rejuvenate. La Fontaine, at the end, leaves the ambassador to which he dedicated his fable, the Athenian orator, politics, war, the devil and everything that goes with him, to write: 'The world is old, they say: I believe it; yet, we must still amuse it like a child.' The same form of the branch returns once more, but human: the old man and the child. The obstacle to the path, the interesting narrative, the flow of my soul, genealogy itself, the son and the father …, do they obey the same schemas?

Narratives

That is the power of fables, the power of narratives: arousing attention, of course, but also metamorphosing the fickle people into combatants. Where does the miracle come from? A cock-and-bull story stirs up a population and brings it to the walls before the enemy. Plato too evaluated the force of rhetoric and the influence of the sophists on politics with distrust. Are we unaware today of the omnipotence without any countervailing power of the media, sounds and images? When I was relating the Grand Narrative, the only attention I was paying was to its objective truth. The latter is my passion but doesn't carry away very many people. Its power flows less from its truth than from this passion-arousing capacity on the part of narrative. La Fontaine evaluated it by recounting that an orator transformed the listeners' attention by himself recounting the ramification I have described: in the spot where a stem breaks off, he placed a graft. He grafted. The strange power that creates miracles lies in the branch. Narratives stage it. How? At least by speaking.

A vain people think the prophecies of the Bible, the parables of the New Testament, the Acts of the Apostles or the Epistles of Saint Paul are full of sermonizing. Not at all: history and old wives' tales are recounted there endlessly. Inexhaustible, Stephen repeated the genealogy, father and son, of the family of David before dying under the rocks; from synagogues to ports, Saint Paul recounted the Resurrection … The aoidoses, the Homeric poets, the griots recited. Montaigne: 'Others form man; I recite him.'[4] Without stories, there can be no culture; there can be no culture without literature, whether popular or distinguished; there can be no religion without a narrative. It alone converts; it alone transforms groups and persons. Speech creates.

The ramification of languages

Plosives, fricatives, labiodentals …, consonants constrict the emission of vocalises or vowels in their passing. Lips, tongue, teeth and palate form complicated barriers, closures or baffles …, whose arrangement incessantly breaks, intersects and reorients the vocal flow. The voice crosses difficult apertures that articulate it. It stumbles with every obstacle, negotiates them and, in getting around them, finds itself swerved by this. If it didn't recount any deviation, it would only utter hootings. Without any change of sense, there can be no sense. Even the vocalises of music break a continuity by means of inflections. Incessantly bifid, tongues multiply branches.

Sculpting, in the mouth, articulated oral language in this way, the couple consonant-vowel repeats the fable of Ceres's trip and the river encountered by the animals and goddess. Flying or swimming, the two brutes accompanying her cross the river without any difficulty, while the goddess bumps into a problem: animals don't dispose of this language that causes gods and men to stumble along exquisitely serrated channels … They whistle, bay or bell … Ceres bumps into the bank the way voice hits the teeth. Are our languages articulated by flow and barrier just like the path and the river, just like narratives, just like the time of our consciousness, just like our lives and our works, just like the world …? Does the ramification carried out by an event on a format unite – oh, marvel – the signifier and the signified? Does the medium of the message imitate the content?

Do you seek to recount? Tell how language itself speaks. The form furnishes the content. Every story in the world lies in lexicon and grammar, in consonants and vowels. My entire philosophy cries out in letters and voice. La Fontaine sang of the power of fables; I celebrate and cultivate the power of language. To each language, its branch.

The ramification of consciousness and of desire

Better. Why do the people, flitting from thing to thing, listen with bated breath when the ford brings the goddess to a halt? *L'Arlésienne*'s storyteller asks: where, yesterday evening, did our story stop? And the child answers: at 'and then' …[5] You were talking, and then went silent; the barrier of the night formed an obstacle to the flow of your story. We left off at the suspension of desire. A river of sleep made the soul's travelling drowsy, … and then … the soul woke up …, and then the Bosporus's cork burst, and then *Homo sapiens* emerged, and then Saint Francis took his clothes off … Yesterday evening, I went to sleep; upon waking, I found the branch again: and then … Why does this suspense cause such a festival in my soul?

Not having any golden bough or branch to illuminate the mysteries, like Dante's bough, which led him through the *Divine Comedy*, I don't know the *self*, so named by others. I have never known how to descend into this Underworld. But I feel it, and I feel myself carried away as it were by what was formerly called the flow of consciousness. The internal consciousness of time coincides with the time of consciousness, and this latter no doubt coincides with time in and of itself. Rousseau let himself go, fluctuating, lying on the flat bottom of a bark on Lake Bienne; Lamartine suspended its duration on Lake Annecy; from on top of Mirabeau Bridge, Apollinaire watched the Seine flow; Bergson evoked the stream of consciousness; Whitehead described the flowing that passes; *Geometry* (p. xxxiv) analysed the expression: it does *not pass* [*cela ne* passe pas], in which the same word is repeated, with the verb expressing the flow and the adverb its stoppage. The foot advances a step [*pas*], but rises, motionless, stunned and taken aback, as though to deny it.[6] The eel and the swallow pass across, but Ceres does not pass across [*ne passe pas*]: and then … Time and consciousness mix the obstacle to the step with the obstacle-free passage: the suspense of the narrative, the suspension of desire. Does desire ramify like language? Who will ever know which one starts things off? Does narrative ramify like desire, desire like language, the latter like consciousness and all of them together like time? And then …

Branch-time

Many linguists give the Latin *tempus* two opposing Greek roots, the verb τείνω [teinō], to stretch out, continuously … and another one, τέμνω [temno], to cut up or interrupt: again stem and branch. Émile Benveniste thought that the set of its strictly Latin compounds, such as tempering, temperance, temperature, tempest …, which designate a mixture, expressed its meaning more originarily. Thus time, *le temps qui passe*, *Zeit*, would come closer to weather, *le temps qu'il fait*, *wetter*, about which we say that it mixes hot and cold, dry and wet, shadow and light.[7] When we say that time flows [*coule*], we forget that the Latin verb *colare* means filtering a mixture. Thus time percolates more than it flows: it filters. It passes and doesn't pass.

Time mixes the continuousness of tension with the discontinuousness of rupture, a flux that flows with a filter that prevents it from passing. Like narrative, whether grand or small, like desire … The nature of time lasts, whereas a time-counter cuts it up. Thus, a number precedes or follows the interval between it and another number on an endless line or a fragment that's as small as you please; the power of the continuous scatters to infinity as many discrete multiplicities. Like the old Nile, the father-river is broken up into daughter-filter-cataracts. And then … the water no longer passes the way it used to pass. Likewise, a fresh wind is fringed into light squalls; it causes the lateen yards to incline multiply. On the open sea, waves divide up into wavelets. Swaying and diverse, the rolling sways but jostles: the vessel is passing over a pile of rocks. Flames descend split into tongues of fire. And then … everyone speaks the hundred dialects that have diverged since the time we separated. Pulverized by mutations and the deadly filter that is selection, evolution explodes into bouquets, droplets and jets of life. I think from intuitions to deduction, from vivid sights to developments; I love from loves at first sight to long faithfulnesses. And then … I don't think the way one used to think. I sing of flows and inflections. I remember and forget … and then … forgive. Time allows me to understand that I love a little, a lot, passionately, madly and not at all … at the same time. I hear a passing lull in the conversation, but tinnitus murmurs in my ears with legions of jinns. Globally and in the instant, my time wakes and sleeps, flows and breaks up, counts and murmurs, sings and dances, escapes me and persists, by pizzicati and held notes, quarter rests and fermata, vowels and consonants.

Made of time, consciousness mixes the day-to-dayness of custom, traditions, habits and laws, which keep it now similar to itself, with the newnesses that awaken and change it. I live, act and think by exchanging this same and this other, fluctuating on this vernal point in which, at the crossing of two roads, springtime is born. My ramified time is suited to the ramifications of the Grand Narrative.

This suitability founds all knowledge.

Disquietude

In the literal sense of the Latin *in-quies* or the German *Un-ruhe*, disquietude tears one from rest, from the calm of a stable state: it removes from equilibrium. And then … the pendulum of the clock beats, without remaining vertical, from left to right. A ball descends into a pit and, with its momentum, climbs back up one of its slopes, suddenly stops there, falls back down, overshoots the bottom and climbs the facing slope. And then … a spring vibrates: the removal of equilibrium compresses it, accumulates, on this side, a power that abruptly releases the spring on the other. Hidden in the depths of the clock and united with the pendulum's periodical movement, this spring counts time. Time is measured by spring and disquietude.

I meet with some obstacle opposed to my project: pressed against a wall, the flow of consciousness, as though elastic, accumulates energy there which, if it finds a path to get around this obstacle, overcomes it, infiltrates elsewhere, surges up via an unexpected exit, like a new branch, in a different direction than the project diverted by this obstacle. Disquietude produces my spring and my resource, counts my time and gives me its motor. Invention and conversion shoot forth from this resource. Would a newness surge up in my time without alarming me? What risk drags me out of my bed early in the morning?

In the same literal sense, ex-istence drags me out of rest or removes me from equilibrium. Thus ex-istence launches me into time. Rocks and the dead rest, serene. Without disquietude, would I exist? This question does nothing but repeat two equivalent words. I exist means: I am disquieted. I am disquieted, therefore I exist. Descartes's famous *cogito* says it without saying it since *penser* [thinking], in the literal sense again, means weighing: assessing a weight on a balance. There can be no weighing or seesawing without little or big deviations from equilibrium. I exist; therefore my time has this motor. Without this disquietude, I do not exist; without it, my state is equivalent to death; disquietude resurrects. A branch has the forked form of a deviation and the motor function of an accumulator, without either of which there can be neither time nor existence, whether objective or subjective, natural or cultural. By removing ourselves from deadly equilibrium, disquietude causes us to be born.

Gather in and store the treasures of disquietude; it throws one into existence. Without the deadly dangers being run today …, would we know how to transform some event into an advent: would we invent a new world?

Advent

How are we to be reborn? Through good encounter *[encontre]*

'She came through the doorway, and, like lightning, I was struck by love at first sight,' he says. Disquieted, he retracts this: 'Have I really forgotten it?' He nevertheless persists: 'To my recollection, the passion that caused me to worry for so long began that morning, the one with the soft sunlight, light breeze and transparent September mist licking the tree in front of the house. Starting from that encounter *[encontre]*, my life bifurcated.' He returns to his doubts: 'Nothing happened on that morning.' Yet he affirms: 'How can it be denied that there was a before and an after? The more time goes forward, the more importance this moment of nothingness takes on, which no doubt had no importance when it happened.' Great loves begin with neither lightning nor strikes. With its prefix indicating repetition, *rencontre* [encounter] shows to those it causes to be born or reborn that they have forgotten their *encontre*, that is to say, the first time: always virgin at this advent.[1]

At the junction point between the twig and its stem, the event causes bifurcation; the advent, for its part, marks, on this same point, a birth. The event can remain sterile, whereas the advent produces; tearing up a monotonous format, the event emerges as an exception to a rule, a deviation from a habitual equilibrium, an interruption of a sequence; the head of a sequence, for its part, an exit, the advent causes, in addition, an existence to emerge, causes subjects, a history … to be born, soon to be equipped with long laws: production, origin, beginning …, the appearance of this *rameau* [branch], which, in French, has the same root as *racine* [root].

From where do these newnesses gush forth? Are they announced on dove's feet, like a thief in the night?

Autocatalysis

There are no glaciers on the Canadian Shield or across the Siberian tundra. Yet, on the same polar cap, at the same latitudes and therefore in the same climates, Greenland's soil is buried every year under the weight of an ice sheet several thousand metres thick. Why don't the same causes produce the same effects?

This is why: long ago, during a winter, it snowed a little there, at altitude. Very little, barely above equilibrium. But the thin layer deposited in that way did not melt the following summer, cooler than usual. These long cold spells arrived; no one noticed them. The next winter, it again snowed little but enough to cover

over the previous snowfall, already frozen. From the second year on, the crust so formed defied the summer thaws, weak at those high latitudes. This recommenced so that once the thickness of the ice was sufficient, no month of July could ever melt it. So, a self-perpetuating cycle started, which ended up in those gigantic glaciers, which territories that didn't experience such beginnings, almost unobservable, were lacking. Thus Greenland is crushed under dense masses that can't be found in Canada, its neighbour, or in Siberia, farther away. In 1952, Cailleux called this process: autocatalysis.

Some almost invisible ripple on the water, equivalent to one that doesn't announce anything at all, announces, for its part, an emergence. After such an advent, the adventure that begins in this way adds a stratum into the enormity erected in this way, necessary, sweeping everything along with its growth. Thus, newness arrives on a butterfly's wings.

The birth of the Earth and of a group

The Earth, every day, receives thousands of asteroids. Light dust, medium-sized granules …, notable rocks dig amphitheatres on it that are sometimes visible. Repeated five times and at intervals of hundreds of millions of years, a gigantic impact annihilated more than ninety per cent of the living species. Yet, at the beginning, one grain no doubt encountered another grain, which, by themselves, attracted a bigger third one, which, already substantial, exerted a more considerable attraction. Thus a self-perpetuating cycle, whose steady-state regime and its limit, our planet, are known to us, formed this planet by accretion, so much so that it still burns inside with these impacts. Who will lay his or her hands on the first rock?

How does an empire begin? On an area of land, termites put down clay balls, each of them depositing one randomly. It can happen that two carriers might put, here or there, two balls one on top of the other. More voluminous than its neighbours, this mass attracts other termites, who, obligingly as it were, add their ball to this spot rather than elsewhere. The growth of giant turrets ensues. With this termite fable, I once recounted the beginnings of the Roman City and the 'causes' of its imperial greatness (*Rome*, pp. 1–5).

Inclination

Even though, in *The Ladies' Delight*, Zola had described its behaviour, the French language does not, alas, have the word 'serendipity', used in English ever since Horace Walpole cited *The Three Princes of Serendip* (Ceylon), whose story stages the delight of finding what one wasn't looking for. By completely changing, one fine morning, the order of his shelves, Boucicaut in fact made the housewife get lost in the department store he founded, thus transformed into a labyrinth, so that, come in to buy a dress and a basket, she would come back out from it defeated by ten unexpected temptations and therefore overloaded with unanticipated purchases. How many times have you gotten lost in a lexicon, like a lady in Le Bon Marché, running from one chance-encountered word to another, even forgetting the one you were looking for, drawn by the double delight of ignorance and discovery?

Some given idea is often grasped in a strange context. Does this random process lead to invention? Does genius come via serendipity? Provided, of course, that one live day and night attached to the store.

Announcements

At the beginning of every play, a comic or tragic author presents a list of dramatis personae. Given millennia of farces and dramas, guess which character is cited most often: the messenger, whatever name he may carry. Not very talkative, he announces events and therefore renews the narrative's development; he surprises, overturns actions, disrupts equilibriums and destinies by means of crises and peripeteias; he disquiets and, striking the coups de théâtre, brings about advents. The ancients revered him by the name of Hermes. I once proposed the caduceus as the hallmark of philosophy, history and the sciences. In monotheistic religions, he moves in the form of an angel. A parasite sometimes, he disrupts communications, vaccinates organisms, metamorphoses things …[2]

I repeat, so we will continually be astounded by it, that, from amorous encounters to glaciers, from the Roman empire to the big bang …, these beginnings concern matter, living things, history or the emotional realm, subjects or objects …, occur in myths, the sciences, religions, works of art, politics … A stone that causes one to trip or a breath of wind reorients; from the *clinamen*, a tiny disquietude, a world is born; from nothing, an existence ensues: this one or that one?

The principle of reason

When, in science, we pose the question *how*, we answer with the cause; when philosophy asks *why*, it is looking for a reason. Why, precisely, does something exist rather than nothing or this particular thing rather than that one? There is always a reason for it. Whether visible or hidden, its principle makes one believe that reason remains stable and that it can only be conceived to be thus. No, reason varies: from a huge quantity to zero; in size and shape; in quantity, nature and quality; in effectiveness as well, whether positive, zero or negative. The semi-nullification of reason in the preceding examples amounts to the minimum limit of these variations.

We meet with the other case, no doubt the most important one in the world: preformation presupposes that a reason can exist that would integrate the whole of causes so that the sequence of time to come will unfold it and only refer to it: a new extremely rare and limit case. Whatever Laplace's demon and its religious twin, said creationism, may dream about it, does reason attain this paradise of preformation? The chaos of things opposes this. I dream of a God continually blowing unexpected contingency over the ever primal waters ..., more expert than the old programmer: since He already thinks the unfolding of everything, what's the use of creating? So would perfect understanding yield to this stupid repetition?

Seeking reason, our experiments and our knowledge also evaluate its subtle variations, which can go as far as its nullification. This changes the sciences, philosophy, our ideas of the real, of things and of humans. We are entering another cognitive era. This book is continually celebrating the birth, without date, of accepted contingency: this, that could have or might have not been born.

Hence, by way of the promised answer, a repetition of the examples. Why do we reach the beginning so little? With its thickness – that is all we see – the frozen ice sheet weighs on Greenland, drives it down and conceals it; Planck's wall separates the big bang from our physics; Roman greatness suppresses its initial weakness; have I ever been able not to burn with love? The curve of a road blinds, with its angle, the direction preceding the bifurcation and discovers, when one turns around, other predecessors behind the new orientation. Amnesiac, generations, on the other hand, take their newnesses to be steady-state regimes. A consequence quickly folds itself into cause of itself. Everything begins with what is erased by what follows. Inventions dissolve the stiffness they make more flexible. Nascent life conceals the death it takes over from. Every production comes out of a death.

Does every beginning amount to a resurrection? On that spring morning, bearing vases of spices, two women, one of whom was

Mary Magdalene, were tearfully hurrying to the corpse of their hope. Through the inspection hole of the half-open aperture, dazzling men, two angels in white, spoke: he doesn't lie here. *Hic non jacet*. Empty tomb; no witnesses. Nothing, absence. Who will believe these old wives' tales? Billions of humans. Now here is the stem concealed by this new branch: the spices, bands and linens … mark the ancient custom of mummification, abandoned by the narrative. The modern era commences from forgetting corpses. Every birth nihilates a death.

An extract from a poetic art: Exordium

And how are these narratives themselves born? Did writers recognize, in the practice of languages, secrets the sciences took so many decades to discover in the things themselves?

Your words, first: from what night do they rise? If you have a vocation to write or speak, learn primarily how to perfect your exordium. In its advent, the entire aim is condensed, as in a reduced model. Its flash of lightning kindles the eyes; its signal opens the ears; its impetus impels listeners along the path that will make them fall into Demades's trap. Three coups or strikes, a lightning strike of love at first sight [*coup de foudre*], a coup d'état, a coup de théâtre … curtains raisers: without a brief presentation of who and what it is about, there can be no attention, no spectacle; you will not go down well. The ever so current formula of telling jokes at the beginning descends from the *captatio benevolentiae* the ancient rhetors recommended to

Rome. Capture goodwill: captivate, enchant … immediately or never. Dead men, stand up![3]

And since I am speaking Latin, this exordium, whose form entices us, comes from the verb *ordior*, to begin, from which weaving draws *ourdir* [to weave], whose original meaning, a concrete one, designates the gesture that intertwines the thread of the warp and the thread of the weft. Text or fabric: stems and branches.

The birth of subjects

Therefore a custom that was running its course becomes suspended: an incident interrupts it, a disquietude takes shape, the weft crosses the warp; so, to fulfil the expectation issuing from the break – and Ceres, what did she do? – one must indeed recount … The scene opens.

'I was going up to the city, coming from my home, when someone, having recognized me from behind, hailed me from afar with a mocking voice. "Won't you wait for me a moment?" Whereupon I stopped and gave him time to catch up. "Apollodorus," he said …' Plato, it is said, was still working on the incipit of the *Symposium* on his death bed. I was calmly coming from my home, lower down, as usual … There is nothing to recount: a regular line, a quasi-dead stem. Suddenly, at a shout, someone breaks everyday life: Glaucon interrupts Apollodorus's repetitions. Everything gets underway at this fork-point. When, later, the people at the table are discoursing on love, who will remember that someone was leaving to run his errands? A good exordium interrupts an ordinary or formatted movement with a break; a discontinuity ruptures a continuity. Time passes and doesn't pass. It begins all the time: the stem makes way for the twig. Hold high this golden bough or branch, which allows descending into the Underworld. Everyone will follow it.

An advent creates or recreates the subjects themselves. Two friends, for example: the one was leading a life that was so difficult he had lost all trace of the other. Oh surprise, here he is, in front of him, emerging from the wings, like a ghost come back! Everything changes.

'Yes, since I found such a faithful friend again

My fortune is taking on a new face …

… Who would have thought that on a shore so deadly to my eyes

Pylades first would have been presented to the eyes of Orestes?'

Andromache begins with a continuity intersected by an advent that produces two subjects thrown into the action to come by a disquietude. The encounter – new – of the one causes the other – gloomy – to be reborn from a quasi death. Likewise, *Iphigenia*'s incipit opens before the dawn, as the king is waking up his servants… *Phèdre*'s incipit, in its turn, reproduces this schema, which is repeated by the more complex one of *Athaliah*:

'Yes, I come into its temple to adore the Eternal;

I come, following the ancient and solemn custom,

To celebrate with you the famous day

When, on Mount Sinai, the Law was given unto us.'

I arrive: the annual rite, a little event, reproduces the big advent from which the law, long ago, came, a law given unto Moses by Yahweh. This exordium doesn't merely designate the Temple in Jerusalem for the unity of place, the famous day for the unity of time and the celebration for the unity of action … but really creates the personages. Who is speaking? I am. To whom? To you, with you.[4] Who is speaking, once again? We are, all together, pious Jews celebrating the birth of our chosenness. Who spoke in the past? For once, Yahweh Himself, to Moses. Before Moses, we hung about in Egyptian slavery and death. We were born at that time; we are being reborn this morning. *Athaliah's* incipit leaves death behind it.

'We were in study period when the head-master entered, followed by a new student not wearing the school uniform and a school servant carrying a large desk. Those who had been asleep woke up, and everyone rose as though surprised at his work.'

Again, the stem and its bifurcation: Flaubert sets up the formatting of the regime, sleep and work, better yet, the ironic purring of an exercise so stupid that everyone falls asleep at it; then newness bursts out, in three arrivals: entering, waking, rising.[5] The emergence of subjects in the face of the subject that has arrived. But, when Emma was slowly wasting away with love, will the entry of the head-master and the suddenly straightened up indolent students be remembered? Simple events …

Good news being given a variety of forms by their incipits, the Gospels instituted a religion of advent, whose calendar only celebrates beginnings: Annunciation, Visitation, Advent, Nativity

… Circumcision consecrates entry into the chosen people … Resurrection forever postpones the necessary end. Christianity celebrates the omnitude of newness. Is there a more complete figuration of it than a still virgin woman after conception and childbirth? In this 'still virgin', simple and profound, we recognize the artistic masterpiece, intuition, wakefulness, dawn, the act of love … No matter if the universe dates back fifteen billion years, the cosmic microwave background, present everywhere in space, virginally testifies to its beginning.

Thus the following were born in the same way: the universe, the planet, glaciers – the inert; narratives, loves, inventions, institutions – collective, cognitive or intimate; men and women – subjects … The parallel, which surprised but dismayed in the format and delighted in events, is continued in advents. Origins shoot forth the way peripeteias bifurcate; births burst forth the way encounters and circumstances do. We begin as and when we change. Contingency invades circumstances and births in every domain. This ebb of necessity characterizes our world, literally renascent. Necessity, where is your death hidden? Death, where is your necessity concealed?

Up to now, my examples of emergence have only cited past newnesses …, which nothing conceals from us. Where will such newnesses arise from now? From our productions. Technological advances are inspiring the terrors of the year 2000. Do they get away from us? Yes, most often, as we have said. Are we afraid of their destructive effects? Can we predict them? Are we risking the unforeseen? But what do we know of their advent? Let's shed light on our own advent by means of theirs.

The birth of possibilities

Here it is. Bouquets of virtualities shoot up from the new bio- and nanotechnologies: molecules, cells, species ..., humans, relations ... – possibilities. Starting from elements, whether particles or genes, such technologies transform matter and its constitution, life and its genesis, the human and its genealogy, groups ... Is a new reality going to be born? Another nature?

But we think that reality is immutable and single, so much does it serve as our stable reference and guarantee. We inhabit it; our actions apply to it; it lies at the foundation of our behaviour. Prejudged to be lies or utopias, every project that quits reality provokes criticism and laughter. Projected outside of reality, today's deviation causes a newness to surge up, one so global it causes panic, in the etymological sense of totality. We only see monstrosity in it: an intolerable exit from the human and the world, a deviation with respect to nature, ethical offences, forgetting of values, a loss of the divine. So are we going to quit this real, the one that life perceives, that physics experiments on, that metaphysics founds, that morality requires ...? And what if this real, necessary and single to our eyes, amounted to the most opaque of our formats? Justified or exaggerated, these panic reactions discover in us a fundamental attachment to a world from which we don't cast off willingly. We love it like an ancient family house. However pessimistic we may claim to be, it seems, at the moment of leaving it, the best possible world.

How do you know this? What infallible father told you? Once again, wouldn't this real, this world, this humanity amount to formatted customs? Reverse the point of view: instead of judging recent projects from the point of view of venerable philosophies, religions, anthropologies ..., discover to what extent these deviations from humanity and from the world question these cultural formations and discover the contingency of their guarantees. In their way, all these formations, disquieted, ask the question: why knock together vessels to give oneself up to the fortunes of the sea when a solid ground carries our stable wisdoms?

Serene, I sometimes envision the world in which I live to be the best possible world. After all, I have no other but it: a body, an existence, a destinal adventure; not only do I resign myself to these conditions, but I wake up in the morning joyful about launching my life by means of this body, in this world, with others. Without any lies or dreams, I try to practice this lucid wisdom, although actively enough to seek elsewhere solutions to my feeble thought, to the defects of my organs, to the obstacles of my environment. But this lively joy doesn't so much sing the necessary as it does the contingent: I could have not been born; this braid of wisdom and jubilation comes from a weaving of modalities.

Outside of mathematics, there is little necessity. This world could not exist or could do so differently. How would the world survive without the water that a comet, it is said, left on it. A legend relates the strange but ordinary circumstances in which my father met the woman who became my mother. The entirety of the real rolls in the sea swells of contingency. We continually vibrate towards the possible, while tied by the constraints of our impossibilities. That said, we are leaving the house built in this way.

The second nappe of the cone

Here is an old image of these newnesses. At the end of *Theodicy*, the goddess Pallas leads Theodorus, the High Priest, to the final story of the pyramid of worlds: at its extreme point, she reveals to him a hall so beautiful he faints from it; this is, the goddess tells him, after having wakened him, the actual world, our world, the one, the best. Below, in the lower nappe of the volume, you see other halls multiply in infinite bifurcations, possible worlds that God, at the moment of creating, did not choose.

In this sublime description, Leibniz persuades us that God eliminated them because they entailed more evil than this one. I would have loved to follow this visit and smile, at the top, at the stillborn branches where I would have lived as a sailor, a composer of music and not a writer, as delighted and unhappy in different ways … Such branches inspired *Jacques the Fatalist*, in which a thousand possibilities shoot up at every crossroads, in which Diderot continually deviates from the road, capering from unpredictable events to bifurcations, caressing in passing ten virtually ramified beginnings; in rereading it (always virgin), we plunge into the actual world of our sciences and technologies, keeping us in suspense, a world traversed with the fateful and newnesses …

In short, the top of the pyramid, alone real, vanishes, towards the lower floors, into a thousand aborted possible worlds. The fact that Theodorus, enraptured then in ecstasy, had only looked at the bottom of the volume, where these vanishing arrangements became blurry to his sight, without lifting his eyes towards the extreme point of the building astonishes me. Being the good geometer he is – the author uses a polyhedron, I prefer a curved volume – he nonetheless knew that a cone develops two nappes and not only one and therefore knew that at the top of the summit a second nappe begins showing, in its turn, new possible worlds, the very ones we are causing to emerge today starting from this one; our projects surge up in this conic jet … like the sideways leaps of *Jacques the Fatalist*. We are throwing in front of ourselves the possible worlds God abandoned below the only real one, chosen for its excellence; reversing the creative act, we are making the spectrum of the virtual shoot up into fountains that are ready to be born. From natured, we are becoming naturing, and I shall talk to you tomorrow about the plurality of worlds.[6]

The *escence* of humanity

The cone above or branch surges up from the nappe below or branching. The second nappe takes over from the first one, which is partly abandoned. We give rise to other worlds, other geneses, a new humanity. We construct sheaves of possibilities. But what is new in that? We have always quit our homes: animality too little, Africa in the past, caves formerly, antiquity not so long ago, solid ground in order to sail the mobile sea and the turbulence of the air, gathering and hunting the day before yesterday, agriculture yesterday, evolution little by little … Our species *leaves*. This is its destiny without definition, its end without finality, its project without goal, its voyage, no, its wandering, the *escence* of its hominescence. We leave and cause our productions to leave from us; we produce and we autoproduce through this ceaseless movement of leaving. We set sail. The way we left the uterus and our mothers' wombs, the lands of our childhood, our naive beliefs, ten fleeting historical truths, our peasant language …, the way we left the programme so as to learn and the innate so as to acquire …, we are leaving today the unicity of the real, of the world and of humankind for possible other ones.

What use is leaving, again? Are we giving up a bird in the hand for a hollow two in the bush? What use is setting sail? Have we ever stopped detaching our hawsers? The Presocratics abandoned the earth for geometry: dreams! Saint Paul left antiquity for the Resurrection: fantasy! Hanno, Nearchus, Pytheas of Massalia, Vasco da Gama, Columbus, Cartier, Cook and Cousteau embarked: madness! Learning quits a prior and programmed niche, a genetic automatism, for an adventure without any clear promise: a leap into the unknown! Being moored across with two anchors in safe harbour shackles our complacencies. Stemming from bitter waters, life relaunches itself there by the wind of contingency and the risky taste for freedom. To the sailor goes the first *cogito*: no one counts on anyone except on oneself on the open sea. A free human, you will always love the sea.

Appareiller [setting sail], precisely: the Grand Narrative of technologies

So never have we attained a sum of possibilities as complete as today. To show this, I cannot avoid recounting the advent of technologies, millions of years ago, by a process of externalization or exo-Darwinism, which I found above (first part, pp. 50–54) regarding knowledge and invention, entering into the body and escaping from it.

Here is it: instead of hitting the head of a stake with our closed hands, we strike it with a sledgehammer imitating a forearm ending in a fist. Technology substitutes, for the extremity of the upper limb, an external man-made thing that resembles it. In a certain way, the tool leaves it, the bones becoming, by our hands, wood and iron, like an exoskeleton. Likewise, for the lower limbs, the wheel externalizes the quasi-spherical joints of the hips, the knees and the ankles; walking and running were already circulating on cycles; reinventing them as machines liberates them from the locomotive organs. Thus the breast pump falls from the breast; clothes shoot forth from flayed skin; the cap from scalped hair; the pump from the heart and the baby bottle from the feminine chest.

I have long called this leaving: *appareillage* [setting sail]. The free sea, once again! Similar [*pareil*] to the organ it imitates, a device sets sail [*un appareil appareille*] in this way from the organism, extracts itself from it like a boat departing the quay, prow first, for the open sea. So, outside of our bodies, our organs go to seek their fortune in the world. Afterwards, our running or jumping records improved less fast than our exploits with the cart, become bicycles or space rockets. Flowing towards history, human evolution transformed technologies more than organisms. Hence the name I have proposed: exo-Darwinism. *Sapiens* and *faber* were born through the invention of devices whose form quit resemblance to its womb so quickly that we forgot its origin. Tools leave bodies; technologies flow from a 'natural' source.

The detachable artificial

A baby bottle imitates the breast the way a hammer imitates the forearm and fist; similar to the breast, it set sail from it. But, in comparison to the organ, fixed to the body by evolution, it in addition has the advantage of mobility, of availability, of having a detachable character. Non-detachable, a lion's fur forces it to stop running when this fur causes overheating. The climatic conditions of the niche orient selection first towards organisms equipped with this or that fur, but once attached to the skin, it remains there, while waiting for other constraints and another selection. This wait can last a long time. A heavy jacket would have, on the contrary, allowed the wildcat forced to rest to continue the hunt; it would have been enough to take it off so as to potentially put it back on, at leisure, depending on the quickly changing conditions of hot and of cold. Thus Hercules put on the hide of the Nemean monster; thus, Aztec priests wore the skin of the flayed victim – in passing, let's salute death, always present. Both of them could remove them quickly or at their leisure. A technological object wins out over the corresponding organ because it can be detached when its usefulness disappears. The baby bottle is equivalent to a detachable breast.

Inventing tools, we have replaced the functions of our organs with them; we have even often improved these functions, exchanging the fixity of the functions for the mobility of the tools, as well as, globally, evolution for history. Don't perceive technology in the same way anymore; see it as our body in externalized pieces, independently evolving. I said earlier that we were quitting the real and humankind: said devices have been quitting our organisms ever since *Homo faber* emerged. Evolution produces a body that produces a new evolution.

Revisiting death

Please reckon, on the other hand, the time saved by technology in comparison to evolution, in which it would have been necessary to wait – millions of years? – for its process to equip us with organs as effective as tools. Calculate as well the corpses it has spared, mercilessly eliminated by selection if and when organisms lacked such tools. Define technology as a tremendous savings of time and deaths. Once again, an invention leaves death.

What can we do today about this original fate except to manage it carefully? What do aesthetic fashions, ideologies, recent economic or political conditions … matter in comparison to this temporal archaism, in comparison to the process of hominization, to this surging up out of the body on the part of these new branches? Technology accompanies nature, since humankind itself was born, is still born and will be born – *nascor, naturus, natura* – from making things; thus it was born, *faber*, already fabricating, with its own hands, equivalents to its organs. And thus it had already entered into autoevolution.

The egg: Half-alive and half-object

We come to a singular organ: evolution seems to have separated the oviduct of birds into the uterus and the mammary gland of certain viviparous animals in such a way that, in comparison to the two functions of the egg of oviparous animals, the embryo returns to the maternal body while the fixed stock of nourishment changes into a secretion of the breast varying with demand. This allows us to deepen the image of setting sail and to see how the hammer leaves the arm. Conductive, the oviduct in fact produces a double egg, half-object, half-alive: living certainly, carrying a developing organism, but inert, a little, since armoured with the calcium protecting what is going to be born. Through the laying of eggs, oviparous animals have already taken, for millions of years, the path of what I am calling exo-Darwinism: they externalize the ovum, a sort of exosome. Since they produce, halfway of course, but already halfway, something of the objective, did birds precociously invent what

I was tempted to swell the pride of humans with? Besides the egg so reproduced, objectiveness that has already been produced, like nests and pretools …, abounds around their bodies. Has a certain technicity already ensued from oviparity? The winged intelligence of feathered creatures – the seasonal invention of melodies by chickadees and hummingbirds; the knotted interlacings and the fabrication of vices so that the beaks of woodpeckers can crush shells; the long-range guidance of migratory birds; the courtship displays in which dances, colours and music unite in sumptuous operas … – marvels that fascinate ornithologists and to which I have so far only given a behavioural interpretation through the three dimensions of flight, have new light shed on them here. More generally, do oviparous animals already produce something of the objective? The objective, not yet technology, production bordering on reproduction here.

At the beginning, the egg: the origin of almost every living being, whether oviparous or viviparous, should we in addition reckon it to be, as it were, a material origin of technologies?

Exoskeleton and keratinous appendages

Better, by externalizing the egg, a movable half-object, isn't oviparity an advance over the exoskeleton, still attached? Were oviparous vertebrates thus continuing a vast movement on the upstream side of which arthropods, insects and other mollusks ..., ammonites, oysters ..., in short, invertebrates in general, were already secreting chitinous armour or, by the intermediary of the mantle, shells and carapaces, an exoskeleton in general, the way masons would build our houses via another externalization? Have we ever built constructions as exquisitely spiralling as mollusk shells, often so luxurious that we don't know of any culture that doesn't seize upon them to adorn their bodies or use as currency? In short, did technology begin from the Cambrian explosion, that Paleozoic Era in which hard parts appeared? Didn't these hard parts invent the exterior of an interior, a framework and protective walls for the soft and fragile parts, as it were? I dream of this antiquity – a half billion years – for the advent of technology. Truly, without grafting history onto the Grand Narrative, how are we to resolve today's problems in depth?

Later, with vertebrates, feathers, hair, hooves, claws and teeth, the carapaces of chelonians, the scales of the pangolin ... continue this vital flow of proto-technicity ... From the oldest exoskeletons, perceptible starting with the Burgess Shale, all the way up to the appearance of these recent keratinous appendages, this giant chain of intrabiological, primitive, 'natural', still not detachable attempts, announcing future externalization, has carried on. Do technological graveyards (polished stones, ancient ruins, junked cars ...) fall into line with the fossils of the Cambrian?

From oviparity to viviparity, moving back then so as to jump better, as with some instances of neoteny, evolution reverts from the externalized egg back to the interior of the organism with regard to the uterus and the embryo, but, finally with humankind, suddenly returning to this, explodes tremendously with devices. The breast falls into baby bottles, those inert half-eggs. Half-egg: a stock that doesn't carry an embryo; inert and not living, made of clay or glass, the way an egg is surrounded with calcium. Our production bifurcates from reproduction. By means of walls and roofs, we repeat the construction of shells or nests, in infinite number, establish networks and cities, in limitless forms ... The image of setting sail transforms into a kind of laying of dead shells, peculiar to certain viviparous animals. By means of living reproduction, an entire organism resembling its progenitors is replicated 'in flesh and blood'; by means of technological production, the performances of an organism are reproduced, almost in the sense of representation or copy, but analytically, function by function.

Not only, for fetishes and *Statues* reproduce the organism globally as well. Not so long ago, I dreamed that the monsters and chimeras left to us by a certain antiquity, and whose forms united humans and animals, sometimes even mixing them – the Assyrian cherub: old man, eagle and bull; the jackal-headed Egyptian Anubis; or Quetzalcoatl: the feathered Aztec serpent ... – related, in a religious context, the long, difficult, never completely finished passage from the animal to the human, and therefore recounted the process of hominization. Did the civilizations that bequeathed us these fossils sculpt what we call evolution? Since said fetishes statufy its theory, we didn't know how to read time on these arrested images. The process of hominization is deciphered there, pious.

On balance, technology gushes forth from evolution, accompanies the tree of species, appears starting from the shells of invertebrates (starting from sporangia, those vases carrying seeds?), in the reproduction of oviparous animals and certain viviparous ones, then explodes, and already in beavers. Seeming to turn their backs on nature, cultures, productive of devices, therefore don't render us so foreign to this natural arborescence and its multiple branches ... nor so divine for the enthusiasts, nor so satanic for those plunged into mourning. Evolution produces, of itself, technology, non-intentional and programmed at first, therefore making it be born naturally, and then, with us, leaves the program and enters into learning and intention.

This broad view redraws our place in the *élan vital*.

Heat, solids, fluids and negentropy

Up until now, I have seemed to confuse technology with the solid inert: baby bottles, statues ..., bronze, wood or iron. Have I committed Bergson's error, for whom coherence and rigidity defined the intelligence that fabricates? Fluids flow with the same movement. The fire-based machines of the Industrial Revolution transform energy thanks to liquids or gases transporting heat. Have I neglected to warm up the baby bottle that gives milk to the newborn but also the calories 'naturally' provided by the mother's breast? So the entire reasoning above is maintained while including the thermodynamic newnesses already discovered by evolution with warm-blooded animals and homeotherms. Engines function like organisms … although the efficiency is not as good; we externalize the internal environment, its Carnot cycles, its exchanges in deviation from equilibrium. But, before the modern engineer, the Neolithic farmer had already sought to manage the refined mixture of heat, wet and windy necessary for Thermidor and Messidor.

Moving from ordinary energy to negentropy, we are lastly designing software that receive, emit, transmit and translate information 'like' tissues and the nervous system. The telephone externalizes the ear and clamours; the computer detaches itself from the head like the hammer from the fist. The recent tools follow from the same flow, running then towards the cognitive and going back towards the living. After the agricultural technologies, which farm flora and breed fauna, the nano- and biotechnologies return to reproduction, from which I have just departed with the bird oviduct.

The origin of the quasi-object

What relationship do we maintain then with these new products that, impelled by a multimillennial impetus, we cause to be born or externalize? As with the facts of knowledge, described above, we reincorporate them!

I have no carpentry or smithing experience, but I have dredged rocks from the bottoms of rivers, worked the land, piloted vessels and written books. The pitching under the legs dances as with lovemaking; armoured in its shell, does the king scallop take a similar fluctuating part in the surf? Sailors and ploughmen caress, force and penetrate riverbeds and furrows like feminine genitalia by adapting to and obeying them; seeds, words, thoughts in profusion ... secreted ... gush forth, quasi spermatic ... for sowings. Climb to the top of a crane and experience how fast you become a wading bird planted on a long leg, watching over the construction site from the cockpit-eye, taking and leaving, with your long beak fitted onto the handle of a long neck, form panels and concrete ready to be used.

The relationship to a man-made thing quickly retransubstantiates it into a living thing. Quasi animists, we personalize it: jealous, my truck immediately breaks down as soon as it changes drivers; a car driven by several hands doesn't last long and soon falls onto the scrapheap like a whore; I loved my boat like a violent, demanding and sweet mistress. Returning to their corporal source, tools become organs again; don't touch my tools, the artisan insists ... or my computer, the writer demands; don't bump my head! The infant with the baby bottle sucks it and handles it like a breast, even though the bottle set sail from the maternal bosom. Anatomy is overflowing with technological terms the way the technologies were swarming with vital words in the past: a translation of this animist proximity (in its genesis as well as by practice) into language.

Lastly, just as the egg-object transits, in reproduction, between mother and child, so the baby bottle, detachable, already marks out, when someone else takes it and gives it to the newborn, familial relations. This is the origin of the quasi-object ... a peace pipe that passes between hands and lips, a ball, token, coins, words, Host ...

Revisiting the origin of biotechnologies and evolution

Like these practices but with an eminent sophistication, do our biotechnologies simply reconnect with this colossally primitive, organic emergence of all technology? Do they too flow back towards their corporal source? Issuing from the organs ..., welcome is their return to the fold! One must have forgotten this imitation on the part of the hammer or software ... and the selection on the part of breeders to be surprised at genetic engineering to the point of suspecting it of monstrosity. The simplest machines have never bifurcated in any other way than the lever from the elbow; the winch amounts to a joint. What have we fabricated that evolution didn't invent? Having left the living towards the technologized inert, technology finds the living to be technologizable; externalized from the cognitive, it finds artificial intelligence. Lastly, production, in the above sense, returns to reproduction, that is to say, twice towards genesis: by returning towards the origin such as I am describing it and by the decoding of genes.

Still ignorant of this second Grand Narrative, the old theory of animal-machines, the mechanism of present-day reductionists ... immobilize this long duration. They reduce its process to an equivalence, even though it took billions of years for evolution to lead the inert towards the living, then some five hundred million to move from the living to technology and lastly to flow back, today, from technology towards the living and the cognitive. So in the final stage of the process or in a loop, biotechnologies return to the living sources of technicity. Conversely, evolutionary time becomes a slow and gigantic fabricative advent: it forms, contingently, bearers of shells, of fangs, of beaks, of exoskeletons ..., circulations of fluids, of energy and of heat ..., emitters, transmitters and transformers of information ... *Tempus faber* precedes the *Homo* of the same name, who imitates it then, as though he were adopting its gesture; entering into its dance, our ten fingers already understood creative evolution.

Avatars

Issuing ourselves from the earth and from metals, from heat and from information, we work metals and the earth, then fires and fluids, lastly the little energies ... like evolution. We take over responsibility for the works evolution accomplished in us and around us. First: having left the inert, the evolution of living things returns to it, archaically in the case of exoskeletons or dams, but in the final stage of the process, through our usual tools. Second: leaving the living, through externalization, our devices return to it in biotechnologies. So technology can be defined as the originary inert, reinformed by the (time of the) living, and sometimes returning, as in a loop, towards the living itself, which is then reinformed by technology. These returns contribute to showing the contingent and unpredictable character of the advance proper to evolution. Used twice, the verb 'reinform' expresses the importance of the role played here by information: I shall return to this.

Just as the living took up the inert again and, in reinforming it, produced carapaces and shells ..., all of them exquisite artistic forms in which the interior closures, windings and chiralities proper to life can be read, so we would take up again our joints, our heat and our software information, lastly our relations, in order to externalize them, thus imitating the production and reproduction of living things, but in diversifying and generalizing them. Could we lastly, and conversely, know the inert and even the living without technological models, methods and systems? Technology mixes the inert, the living and information variably, all three of them continually returning, as in a cycle, one after the other, according to evolutionary time. The inert, the living and the technological thus become avatars, capable of being metamorphosed into each other, over the course of a colossal time, like three varieties of the same type, a type common to these states. Can it be named?

Conversely, evolution itself, which we are beginning to suspect began starting from the very first inert molecules, which were tried, abandoned, transformed ... (yes, mutating and selected in a way), shapes the totality of the ways things are. So might evolution begin with the big bang? Might it pass then from the exclusive domain of the living to the span of the Grand Narrative, history and even, no doubt, knowledge included? A 'substance' flows in this Narrative, a substance that passes through four states, inert, living, technological, lastly immaterial, which our various metaphysics distinguished, but which duration mixes and shapes.

Numbers, codes, notes

Immaterial or abstract, the fourth state, sometimes named mind and sung of as the ultimate crowning achievement of the productions of the human understanding or of raw matter, or as the first project of a preformative God, according to this or that stubborn division, constitutes, to my eyes, although paradoxically to the eyes of many, the fund common to the inert (particles, atoms, molecules, crystal …), to the living and to the technological, the 'substance' common to the three other states, which are unfolded along evolution and constructed by combination and figures … All things shoot up from notes: to say what flows in the Grand Narrative, cosmologists, physicists and biochemists, recent, unite with the ancient Pythagoreans, who saw numbers everywhere. Words and things are coded as notes or numbers, information bits or pixels. Here has returned, but at the limits, the advent on dove's feet …

The universe of fire, the rocks and the ices of planets, the air and the water, the molecules that replicate themselves, the scales of reptiles and the feathers of birds, the strange charm of women and the native simpleness of children, the sparks of forges, the planes and the computers, the marble statues and the symphonic waves …, crystals, flesh, arts and trades, fine arts, even my desire, even my prayer, even my ecstasy …, all of them are composed of the abstract: superstrings and branes, multidimensional spaces, excitations of the quantum field, the curvature of space, figures and symmetries, elements of information, pseudo-points, numbers, probabilities of presence … Radiation and matter ensue from an immaterial ... Elements link together to form things and world by exchanging information.

So knowledge shoots up from what flows in the Grand Narrative. In sequences, knowledge folds itself on to the temporal chains of the world, is part of them, is extracted from them, extends them and imitates them, plays at constituting or reflecting them. Terminal, initial, arborescent, unexpected, repeated, abandoned, ramified, made up of unpredictable attempts in elementary sequences of information and of numbers, it conforms to things made of numbers and of information. *Cogito*: the abstract productive of the world encounters the abstract that reconstitutes me. *Cogito*: geometric and numerical, my bones rest on geometry and the numbers of things. *Cogito*: the abstract that makes me pass from my light and fragile existence, emotive, rare with intuition, dense with inexistence, to the weighty things of the world passes through the abstract that makes the world pass from nothingness to existence, passes through the abstraction, as well, that makes you pass, yourself, from absence to presence. *Cogito*: dense to the gills with informative sequences, I plunge them into the information sequences of things, of the world and of others.

Without adequations between these two abstractions, the one that constitutes me and the one that forms the world, no science, no knowledge, no language, whether rigorous or intuitive, nor music nor poetry, nor belief nor love ... would be born. Without these adequations, mathematics would not speak universally; we would remain strangers to the world. Thinkable thanks to the immaterial operators that construct them, the community of the four states varies in the Grand Narrative. Along this narrative, traceable *a parte post*, the bifurcations of each state – inert, living, technological … – begin, unpredictable *a parte ante*.

The surprising paradox of the fact that the real is born from the formal is erased the moment one understands that the abstract stands between nothingness and existence and, moving from one to the other, forms a bridge between the two. Has it ever had any other status? Neither the concept nor the circle, nor the concept of circle, of course, exists here and now; but who can say that they don't exist in a certain fashion? Some people even say, and I agree with them, that the circle enjoys a real existence, although not an empirical one. Mathematical entities exist powerfully in this intermediate mode between nothingness and the perceptible. Experiencing the things themselves encounters constraints that testify to their existence; other constraints, even harsher, leading to the rigor of demonstration, persuade the practitioner of the inevitable existence of these entities. This abstract mode of existence allows us, as we know, to explain the real world, the living, technology and sometimes even musical cultures and their beauty, to understand them and give an account of them. It in addition constitutes the fabric of reality of the things encountered by work, perception and behaviour. This abstraction makes them arrive to reality. Infinitely light, the formal gives birth to the real.

Today's technological branch surges up from the sum of this real. By spelling so many alphabets, we recombine their notes at leisure. Thus several possible worlds are being born today, which is what I had promised to demonstrate.

Naturance

Described earlier, genuine knowledge transubstantiates its object; begun as incorporation, it ends, when it invents, with an externalization; another form leaves the body. This productive process imitates the living one of reproduction. The evolution of technological invention repeats it. The coupling of the three descriptions launches *Homo sapiens* into a continuous impetus of life, cognition and practice. Transformative, the motor of metamorphoses, it produces boys and girls, thoughts and signs, tools and machines … new … all of them launched into contingency. The world and the human, naturing, are crying out in the labour pains and joy of childbirth.

Today

How are we to move to today's sum?

To the eyes of the person crossing the giant bridge that spans, from Antirrio to Rio, the Gulf of Corinth, above cities swallowed up by ancient earthquakes, the Venetian lighthouse and castle, below on the coast, seem to be dollhouses. Our technologies change scales. During my father's generation, a person might embark in Le Havre on board a sailing ship laden with three hundred tons of cement destined for the ruins of the San Francisco earthquake; our ships have a capacity of a million tons. Found in ice core samples in Greenland, the effluents of the Bronze Age didn't dirty the air of its time much; millions of cars darken the atmosphere of Los Angeles. With demography changing scales as well, billions of inhabitants overpopulate China and India. The internet in space, the atomic bomb for energy, the greenhouse effect for the atmosphere …: since recently, we have been producing objects for which at least one of their dimensions grows all the way up to the dimension that corresponds to it in the world; formerly compatible with the dimensions of the body, technologies have become globalized: the planet is rocking from it. The opposite of these world-objects, the alliance of nano- and biotechnologies with artificial intelligence sculpts tiny tools, atomic sized. Born from these rapid progressions towards the immense and the dwarfish, does the globality I was talking about above in order to understand today's real exceed us because the quantitative all by itself changes the very nature of the posed and resolved problems? Not only.

This total transformation begins, as you will remember, with three geneses: the productive codes of molecules of matter; the genetic code of living things; codes or pixels of information. Knowing their alphabets, we produce open sheaves of possibilities by combining their elements. Yet the knowledge of letters doesn't imply knowledge of the texts they can generate; music scores cannot be deduced from the sol-fa. Have we lost all mastery of our productions?

The sham master

Having arrived at this point, I dared to deploy a technology narrative. A few ideologies accompanied technology's progress. Pragmatism, among them, celebrates the hand and the man-made objects that leave it. We think we master these gestures and their works. Wear and tear and breakdowns, however, inform us that the made object preserves an objective existence, independent of us. It participates in a real that exceeds us.

Dancing as though in lovemaking, my vessel, as you will remember, took on the autonomy of a woman. Corn was planted beyond Mexico; we had to tame the horse on more continents than the person who conquered it foresaw. Invented to prudently limit a genetic modification to one generation, terminator technology allows seed companies to enslave farmers. The *Odyssey* cries out even more loudly, longer and better than Homer's declaiming mouth. How many works and results were diverted from their destination in this way? Published, a given text goes out to seek its fortune in the world, like the broom of the sorcerer's apprentice; false interpretations or unexpected translations will be overabundant. Pleyel didn't foresee Debussy or Fauré. Designed in order to listen to opera in the living room, the telephone is now used for everything except for its initial finality. Hard or soft, have we ever mastered what we fabricated? Even here, we find contingency; even here, the father loses.

Thinking he dominates his productions, *Homo faber* is surprised when they end up at strange consequences, scandalized at the fact that the things that have come from his hands with some given intention would turn in any other direction than the one foreseen: what an illusion of finality! He never operates them by remote control so easily; he masters them here and sometimes; but at times and there, they dominate not only their author but the environment. You will have trouble finding a single tool whose future flowed in the channel its designer foresaw. The lesson of the sorcerer's apprentice lies in the ordinary behaviour of the things we fabricate. Their use exceeds us. We always invent a little or a lot of automaticity. The sapwood of the handle I carve is composed of a real that reacts according to its laws and not my own. The page resists the writer like a stubborn child. Who can claim to be the full cause of the entirety of the productions-effects of his or her will? This independence even reaches less technological projects: appalling, the ways humanitarian institutions have been diverted; ridiculous, those who take fright at the idols they have sculpted. Mathematics goes to the peak of this paradox; coming armed out of Thales's, Gauss's or Poincaré's brains, geometry nonetheless compelled them to think what it imposed; better, it objectively expresses vast sections of the real. The producer's causality weakens.

This can be verified for God Himself: Eve and Adam escaped their shaper; through the sin and the expulsion from paradise, the couple ran towards their freedom. In other words, creation doesn't necessarily imply preformation. The most improbable of the theses set forth by what is called creationism programmes the behaviour of creatures *a priori*. If you so desire, keep creation, but abandon preformationism.

Consequently, celebrating the possession of nature lapses into deceitful publicity; the mastery of mastery, whose formula I once proposed, attempted to manage these illusions. One must never have held a tool in one's hand to think that a tool is always fabricated for one use and one use alone, clearly conceived, entirely subservient. If a lever is used to lift, I'm not able to say what; set up a computer: for whom and how will it be used? This illusion of finality, crowned, askew, by Cartesian mastery or pragmatism, dates back to the carving of the first stones;

let's abandon these utopias. The technology narrative therefore participates in the same contingency as evolution.

In fact, we are little by little exploring what our fabrications can do. Sometimes the artisan, like the artist, works without finality.

Evolution produces a producer of evolution

Descended from life, the technologies return to it, as I have believed I could say. Bio- and nanotechnologies invent composite structures, freely imitating the non-finality of the living and the non-finality of intelligence, whose unfolded spectrum of possibilities isn't known to those who promote these technologies. So we are accepting today what we have accepted ever since the dawn, an uneven, weak and often worthless mastery of our productions, varying like the principle of reason. Apart from the scale, nothing new under the sun. If our terrors, ordinary today, are due less to risks, whose idea presupposes an insufficient mastery, than to this non-finality of the new man-made things, we can put these anxieties into perspective with the idea that past cultures didn't master the pseudo-finalities of the ancients any better. The size changes, not the unforeseenness. Ourselves contingent, we fabricate the contingent; evolution produces us as producers of an evolution.

When we poorly reconcile God's omniscience with his creatures' freedom, we are copying this finalism more than the act of fabrication. But, the more our knowledge advances, the more we take note of our contingency, the contingency of the world and of our actions. Except in cases of extreme simplicity, we have made and will make the unforeseen. The more we approach the Creator, the less we imitate providence. We don't preform anything or anyone: the best teachers don't educate parrots, but autonomous human beings. Children disobey: *felix culpa*; this blessed fault allows them to sometimes get around the obstacles accumulated by our formats. Omniscience would preform someone rather less than half-clever. Let me stress the definition of our new era: contingent evolution produces a producer of contingent evolution.

The ethics of the helm [*gouvernail*]: Precaution and prudence

From this half-mastery are born our anxieties, in the face of which we establish the principle of precaution; it decides beforehand. If a principle, preceding every beginning, remains, it risks strangling the work. Preformationist, it becomes a pretext for motionlessness, a kind of lazy sophism. Let's abandon this prin- and this pre-, false and useless. Since everything moves and is negotiated in contingency, it would be better for this principle to vary the way mastery and the principle of reason do.

So let's invent an ethics in the cybernetic mode. Let's steer [*gouvernons*] productions whose behaviour we never decide once and for all and before all. Following the teachings obtained over the course of their evolution, let's inflect our decisions in real time by practising the prudence of the pilot. At the helm, he steers the vessel following his intentions or those of the collectivity whose plan he is executing, but while continually taking into account the reactions of the swell, the wind, the stability of the ship, its dance with the waves, the mood of the crew, the age of the captain …; headstrong, he holds course without dawdling towards the four winds of the compass rose but changes bearing if necessary, puts into port, heaves to or lies to, retraces his steps, circumvents cyclones and the latitudes of calm weather …; in short, he steers. Aware of the contingency of the world, prudence acts according to the logic of modalities.

So scientific and technological behaviour concerning things resembles the behaviour politicians say they follow with regard to society, composed of Eves, Adams and unruly children. Continually tied together in this book, the connection between the world and humans is continued in morality. We will practise the political and objective ethics projected by *The Natural Contract*. We will steer the planet and humanity by virtue of a single virtue. Mixed with swell and course, the real responds to the prompting of the pilot, who carves his bearing, contingent and necessary, along a possible course that circumvents impossibilities.

Death on the horizon

Praised in this way, prudence envisions the eventuality of shipwreck: none of the newnesses will do. Can we, today, resign ourselves to failure? The globality of our engagements, the projects of other humans and even of different natures ... are opposed to this resignation because we put in danger not only some given local existences but the entirety of the conditions for survival. Once more, our formats are heading towards death, a death that's global for the first time. We find ourselves forced to find an exit hole from this deadly outcome, we who are still condemned to resurrect. I shall attempt to answer the questions concerning this serious steering later on.

The fact remains that every beginning surges up from an aged, exhausted, dying stem; that each of our resurgences gave rise to its branch from this inevitability. However far back in time we may go, every bifurcation of the Grand Narrative opens up a similar exit hole: the Universe expands from a theoretical point, unthinkable by science; life emerges from the inert; multicellular organisms are born from the first bacteria; a species mutates from a less adapted preceding one; culture, in its turn, is extracted from organic life; the human leaves animality little by little ... These bifurcations occur as exits from old places, where, if it remained there, the new branch would die. In the longer and better established time, everything that exists, ourselves included, continually surges up from a prior format in which death would have defeated it.

When we were ignorant of the Grand Narrative and the sequence of its surgings, we would define ourselves as 'being-towards-death'. But our knowledge of nature cries out that, like nature, we never cease freeing ourselves from death. All existence comes out of a nothingness. With the last of the exits out of mortal constraints – inert ones, animal ones … – humanity was born. Accepting this truth, millions of years old but renewed every morning, philosophy, taking over the reins, announces this exit from death towards life. Define humankind as hope for life.

Leaving a place

Producing and produced, projective and prudent, we all left or came out of the earth, and each of us left the uterus. The dawn delivers us from sleep, love from torpor and invention from dogmas. Depending on our wills or desires, especially so that our thought may fly, we leave customs, sometimes, and, today, I hope, we are leaving deadly human history, the state that makes war, the war that makes states, the genealogy of blood and soil, theatrical and deadly social statuses, the libido of belongingness … The monotheists say: the way Abraham quit his country and his father's house; the Jews say: the way the chosen people left Egypt; the Romans said: the way Aeneas escaped from the Trojan hell; Saint Paul: the way the new man is released from the Law, the way Jesus Christ resurrected from the tomb, thenceforth empty. Life abandons its walls, its cities and its barrels. Inventive thought leaves formats. We deviate to exist, a deviation from the stable, from the world, from humankind itself. We continually deviate from these things and this morning as well. The Grand Narrative only recounts deviations and exits. Things leave us the way we leave things.

Leave from where? From a belongingness, from a place. To the litany of newborns corresponds the litany of places: Eve and Adam quit the first garden; the universe surges up from nothingness, behind Planck's wall; Moses leaves Egypt; the mutant bifurcates from its species; Jesus resurrects from the tomb, a place-image for every place because every place in this world marks the place of a tomb; cultures abandon nature; science often contradicts everyday perception; and, once again, freedom abandons the land, the *pagus*, the plot of alfalfa, region, country ..., places where fathers buried their fathers, funeral spaces; the hominian quits the environment, which other living things make their niche, but also that extra place we designate by law, traditions, ethnology, political collective ...; we leave massacre through sacrifice; lastly we leave the sacred consecrated by sacrifice through saintliness ... Death incessantly lines up, in procession, this set of particular and local places ... What is death? What takes a place: here lies. Leave from where? From place. Tombs mark sites. Death marks belongingness and its deadly libido: belonging, dying. What is a place? A site marked by a tomb.

Vibrating, chomping at the bit, being born ..., life and thought dis-place themselves. The intuitive and the inventive free themselves from the site. An old measurement of the Earth, geometry quit surveying and invented another space. A homeless fire without hearth, flying, love absents itself from places and transcends them (*Rome*, pp. 165–6). Place marks hatred, which doesn't leave place. Today, we are quitting the places of the world and of humans, the complete set of our belongingnesses, the sum of deaths. A fundamental format: place and site; their content: hatred and violence. We still have to leave death.

Interlude: 1944–2004

A few days after the landing of the Allies in Normandy, while the battle, on land and on sea, was multiplying military corpses but above all civilian ones, a Luftwaffe Messerschmitt, piloted by a captain who was already famous for his victories during the First World War, was pursuing a twin-engine aircraft of the Royal Air Force cruising in the vicinity, with, on board, a twenty-year-old contender, a champion of aerial acrobatics. A fierce duel ensued, experience against youthful ardour, in which the two aces deployed twenty feats of skill, courage and tenacity. The fight to the death ended with the almost simultaneous explosion of the two aircraft. At the edge of the two blazing clouds, two parachutes opened, and the adversaries landed, by chance, on the same cramped rock, many yards from shore. Squeezed up against each other, the German, who spoke French from four years of occupation, queried the young Englishman, who had just learned French, about the horsepower of his engine and its maximum speed for turning; from his side, the young pilot plied his grey-haired colleague with questions about the old airfoils and the ailerons of yesteryear; a burning conversation began about nosedives, flares, vertical climbs and shooting angles, and especially the new V-2 rocket engines. Passionate about mechanics, they had the pleasure of squabbling in this way, an ancient practice against new things.

At dusk, the low tide opened up dry land. Continuing their excited exchange about propellers, fuselages and cabins, they sought a refuge. Groping about in the dark woods, they headed, in the black of night, in a direction from which they seemed to hear cries and clamours only to come across, flabbergasted, an isolated farm in the forest, where, despite the circumstances, a wedding was being celebrated. Bloody, dusty, smeared with grease, the phantoms appeared before a table loaded with lobster and bottles, set up in the open air in front of the barn, amid torches and lanterns. Fearful and hospitable, the farmers of the building offered the duelists a jet of water in the wash house, two towels and some Marseilles soap, cider, Pont-l'Évêque and calvados; the peasant women had them dance to the sound of an accordion, amid the laughter of the delighted girls. At noon on the following day, they woke up from a memorable booze-up, dazed.

For several decades, the two enemies' friendship experienced not a single shadow.

Reprise: Concordance

How are we to leave death? I'm trying to answer the question and will end this book by going back over its second intention. How does it happen, again and lastly, that narratives and emergences, advents and exits suit the sciences, whether astrophysics or biochemistry, just as well as they do the arts, religions, history, everyday life and surprises of love ... without distinction, whereas our formats carefully separate these cultural formations, which, in distant institutions, form segregated groups?

Global, might the new concordance announce itself, by chance, at the moment when we need it, faced with the worldwide globality of our productions? This concordance began when the hard sciences no longer limited themselves to asking the question *how*, a question defined by positivism, to the exclusion of the question *why*, and no longer considered their objects solely from the point of view of the laws of their functioning. Starting with the nineteenth century regarding the Earth and fossils, the question *when* or *from when* arose. Gradually, all these sciences also considered their objects to be memories buried beneath other codes than those of language; these sciences deciphered the age of the universe, the planets, living things ... starting from archives, which were silent of course but newly readable and often including traces of origin. When positivism separated the sciences from myth and metaphysics, it was opposing the operational explanations of an exact and falsifiable knowledge to gods and nature. Said nature seemed to them and us, with good reason, to be an empty fetish; yet the same word describes, literally and wonderfully, the process of birth and development now followed by the majority of knowledge. From the big bang to the Cambrian explosion, from the accretion of the Earth to the first humanoids, nearly all of the contemporary sciences climb back to the advent of their objects. Hence their unexpected connection – concordat and discordance – with the grand narratives of religions and the little avatars of history.

But, as soon as the 'naturing' question *from when* finds answers and as soon as it unites with the question *how*, which clearly doesn't stop being asked, the sciences cannot help but to meet with the question *what* or *who*, I mean the question of the individual quality of the object that was born in this way and during those times, another way of saying its 'nature'. Over the course of the Grand Narrative, some constant or some planet … comes about with its own characteristics, as well as some species mutated or adapted like so …, not that one, but this one. As a result, a new question becomes added to the first ones: *why this one and not some other?* So, formerly given over to necessity, the sciences discover contingency. The Universe, the Earth, the living things, humanity itself could have not been born or could have been born and developed differently. Formerly leaning on ontology and the common logics, they are now on the terrain of modal logics, in which possibilities and virtuality play. Owing to contingency, the question *why* comes back, a question dismissed by the science of our fathers: *why this universe and not another, this living thing and not that one?* Formerly abandoned in favour of the general, the individual comes back in some way: *who* or *which other?* For other possibilities could have been born. Why this individual detail rather than that one?

On balance, there is no longer one nature, but several; there is no longer this, but a multiplicity of possible individuals. We move from nature natured – the things born, therefore inevitable – to nature naturing – the virtual to be born. Hence the new question: *why not, consequently, cause this or that of these worlds to be born, this or that of those non-real molecules, this or that of these non-born living things …, which of these unknown humans?* Returning in this way, the question *why*, formerly repudiated for its finality, is not so much asked for the finality of a real (a real now assumed to be contingent) as for the intentions that direct our own decisions regarding the possible. To what end do you want to, when and if you can, fabricate this other? The new producer of evolution's global question.

New politics?

From which this concordance is born: just as, owing to rediscovered time, the Grand Narrative linked the sciences with religions and history, this contingent choice of possibilities links hard knowledge, matured, made wiser by these new questions, with novels, literature and law. Scientists create the real the way novelists, journalists, poets, jurists or philosophers do; a new concordat ties these actions to morality and to politics. With the imaginary entering the factual, the tragic will arise: in the spectrum of possible births, the shadow of death looms up. If we can promote advents, how are we to negotiate the ends that, no doubt, will follow them? If we help nature to be born, will we master denaturation? With the scientists of every country not being enough anymore, everyone can and must answer these latest questions, religious people as well as jurists, experts of every culture neither more nor less than the ignorant from every part of the world. Among the accessible possibilities, what new world do we want, all of us, what living things, what humans?

This is why I quoted Saint Paul, who, like us, saw a world die, anxious to promote a new one; by forming, as here, a global thought of advent, he invented a subject suited to defeating death. Disquieted about the same questions, we make every cultural formation contribute to founding again a global subject, a different society, a cognitive, active, responsible one, in brief, a politics. Tomorrow, we'll have to deal with the topic of power.

Three reasons for the concordat

Three reasons, lastly, contribute to the same concordance. I have said the first one, ontological: we fabricate possible humans and possible worlds by means of codes ... notes, numbers, atoms, letters, genes ...; their type is of no import, for having similar gestures. The second one has to do with methods and objectivity. The social sciences readily recount that the objective sciences were constructed over the course of a contingent time, inside collectives traversed with political conflicts and made up of capricious individuals delivered up to the circumstances of events; granted, how are we to live and think differently? Does this dissolve their objectivity? Would plunging a rigorous system into a contingent history of subjective individuals and various singular politics relativize its results? Would a tension exist between the constant laws and the fluctuating history, instances and generalities? Can this tension be reduced?

Yes. Even if the hard sciences proposed ten methods for analysing temporal contents that up to now have been peculiar to the social sciences, these latter would continue to set their historical perspective in opposition to a rigour that would never understand or could never compel the margin of advents, of events, of vagaries, of unpredictabilities ..., details and singularities that are swarming in political affairs. But this supposed inflexibility often reduces to a ready-made and, worse, extremely old idea that the practitioners of the soft sciences have created for themselves regarding the sciences now falsely said to be hard.

For these latter sciences, I repeat, have invaded ten temporal domains: they date their objects better than the historical sciences; giving rise to the Grand Narrative, they unfold a duration that no other narrative could have conceived; their new flexibility includes contingency and change; they give rise to the individual, the instance, detail and landscape: they find themselves on the terrain of historicity. The four concepts of modal logic – impossible, necessary/possible, contingent – used to set these knowledges in opposition two to two; these concepts now unite them. What the social sciences called history often becomes science, and what the hard sciences named science sometimes becomes history.

Hence this book and its main image: the trunk or stem of the arborescence represents 'hardness'; always effective, necessary even, the format carries the exactness of measurements and laws. No science, no education, no work does without it. But, surging up from it, freed from its exclusions, liberated from the group it can federate, the branch, pointed and piercing, causes the unpredictable to surge up, in which newness appears, in which, in addition, and I shall end with this point that seems to me ought to resolve our difficulties, in which, as I was saying, the individual is substituted for the schema, in which the landscape makes the geographic map grow green again ... This is the place, new twice over, where the two sciences progressively join. From their overlapping can be deduced, in return as it were, the pluralist politics of decision evoked above.

Change of knowledge

The third reason therefore has to do with the cognitive: partly thanks to computers, knowledge is changing. The procedural is sometimes substituted for the declarative and at least supplements it; algorithmic thought sometimes replaces conceptual thought and counterbalances it, at least.

Let's define the words of this enigmatic sentence clearly: the declarative or conceptual invents ideas, defines them distinctly, follows the principle of reason in its fixed version ... by following the sequence of causes and effects. The algorithmic or procedural constructs events and singularities step by step, enters into detail, along sequences of circumstances and time. Abstract, the former demonstrates; individual, the latter recounts. An example: with the Grand Narrative, which sums them up and makes their objects come about as dated events, the exact sciences, up until yesterday exclusively given over to declarative thought and concepts, unfold an arborescent algorithm whose step-by-step development moves from format to advent or from stem to branches: in describing these two elements of arborescence, these pages, bifid, graft the procedural onto the declarative.

Up until recently, we didn't know how to measure, calculate, think ... without a concept, without its definition and its field of application, extension and comprehension. An abstract and empty form, we fill it with perceptible intuitions; clear and general, it comprehends a hundred particular instances, whose shadow-stained details we have planed down. The word and the idea of circle, for example, grasp at a blow, under their perfect light, every lopsided ball and every scalene ring encountered in everyday life. The tangible benefits of this declarative: illumination, economy of thought, lightning-fast memory. Philosophy and sciences, in sum the Western cognitive, were born, in the direction of Athens, from this Platonic idea. Who can do without its format?

Yet, ever since we have been working on screens, we have left a part of the work of statement, verification and memory to machines. Rapid, the electronic navigates as many balls as you may like and sweeps across them in such a way that the mnemonic economy of the concept of circle yields less benefit. We free ourselves in part from this formal belongingness. The old image of light passes from clarity to speed: to understand thousands of examples, we have less need for a concept, whose ultra-economical memory we leave a little. Inscribed in the machine, a thousand algorithmic procedures permit us to construct and directly envisage the wealth and details of singularities, consequently not planed down. Abandoned by the declarative, the individual resurrects, overloaded with modes and circumstances, roaming a thousand events, astonished at new things, endowed with a new universality; in this procedural book, the pages abound in narratives and persons.

We stroll around in deviation from the old format. Formerly and recently, we saw, in the telescope, singularities distributed over the conceptual background of our declarations, distant stars over the nocturnal firmament; we now consider the concept instead to be poor shell tossed about in a dense landscape. In the preface to *Paysages des sciences* (pp. I–LXX), I have already described in detail this new entry of detail into the old maps, smooth and empty, the revenge of portraits on schemas, in brief, the ramifications of contemporary knowledge towards a thousand leafy sites. In comparing, for example, the old images of the San Andreas Fault and the images available today, a simple geometric line disappears in favour of a bushy thicket of multiple faultlets; anatomy textbooks transit from abstract schemas to MRI-individuals in which the hip of this young girl or the shoulder of that old man appears; the cosmography textbooks forget the simplistic maps of the sky for photographs of each planet, at some latitude, or of some galactic collision … Certain biotechnologies prepare treatments that recognize the patient so as to cure his or her specific illness. The coloured concrete surges up from the grey abstract; the multiple escapes from the one like a swarm of doves from a magician's hat; life resurrects from the format.

The concept forms a box, whose word says wood: touch its hardness. A porous pocket, elastic, the singularity, less exclusive, slackens, pierced, mixed, tiger-striped and zebra-striped. The concept pulls in the direction of geometry; the singularity towards topology; the one solid; the other flexible to the point of fluidity. A box never comprehends all boxes … If there exists one way to put several of them into one, you will rarely find two …, hence exclusion, forcing …, whereas you can stuff three, ten, a hundred sacks into any sack, crumpled. The old science practices rigid boxes; the new one limp sacks. From the declarative stem surge up, slender and multiple, procedural branches.

Three formats-cities

Let's go back to the images of the old symbolic cities: studying in Jerusalem, passing through Athens and dead in Rome, Saint Paul travelled in the triangle drawn by these three formats-cities of antiquity. Our contemporary ears hear the Semitic holy city resound with the biblical grand narrative, already algorithmic and prone to individuals and their singular procedures in its narratives. Conversely, when we understand, we do so, still today, after the example of Athens, given over to the declarative. When Rome coded its Twelve Tables, where is it placed in this game?

Like his person and his travels, the Epistles of Saint Paul set up a new equilibrium, fragile, between Athens and Jerusalem, conceptual thought and algorithmic unfolding. In a new literary form, partly individual autobiography, he invented concepts in order to understand the advent that had recreated his life, and recounted the advent of another survival, the resurrection of Jesus Christ, a proper name that unites Hebrew and Greek, the declarative and the algorithm. Today, we are living through another act of this cognitive alliance, the first act of which was born in the person and works of Saint Paul, which are half-Jewish, half-Greek, half-event-oriented, half-theological, in which the conceptual contributes to the narrative, in which the narrative aids the notion.

Just as the author of the Epistles no doubt died in Rome, Christianity settled into and would live there; like Saint Paul, should it leave Jerusalem for circumstantial reasons, it will have difficulty rejoining Athens. Concepts in the Greek style are further removed from the Semitic algorithms of narratives and events than the format of Roman law, whose individual and jurisprudential cases would give birth, later on, to the algebra of the Renaissance; better yet, the legal subject, an empty and formal singularity, contributed to the birthing of the Pauline subject the way Roman adoption aided the erasure of the genealogy of blood. From the cognitive point of view, Rome is closer to Jerusalem than to Athens.

For four thousand years, Western philosophy, theology and sciences had descended from Athens, the fount of concepts, and not from the two other cities, sources, for their part, of histories, narratives, jurisprudential cases and algorithmic sequences. The cities don't understand one another. Trained from my youth in concepts in the Greek style, I failed my entire life to understand the advents and singularities of narrative religions, as well as, more recently, to assess the newness of the Grand Narrative and the surging up of its branches. My conceptual lights left the algorithms stemming from Rome or Jerusalem in the dark. Our contemporary knowledge is finally grasping them and unfolding their cognitive riches. Formerly indisputably in front, Athens is regressing. In our new cognition, the three tributaries flow together; here we find the three cities equal to each other, at the moment of melting into the universal.

Another interlude

Between the first century and today, Leibniz and Pascal had already practised a kind of Pauline equilibrium between the invention of concepts and narratives of singularities. Both of them inventors of the first calculating machine, authors of infinitesimal algorithms, drafters of procedures such as the arithmetical or harmonic triangle, both of them, tirelessly, recounted: the one related Martin Guerre or the Polish twins, which Christiane Frémont shed light on wonderfully, the other diverts us by describing the solitary in his chamber or the balance artist on his plank … Constructing procedures that make a singular individual into an incarnate universal, they indicated the contemporary direction towards the synthesis between universal mathematics and the metaphysics of the individual – the synthesis-source of the new concordance.

Saint Paul announced the first concordance; Leibniz and Pascal prepared ours. When I, a lost individual, wander in the world, hesitant and step by step, by leaving these three cities for the universal, I recount procedurally; thus I spoke of Saint Paul.

When I evoke the algorithm as the set of possible narratives, it becomes true that I declare. Through these intersecting approaches, this book attempts to tie, again and for today, universal mathematics, stem-father, to the metaphysics of the individual, branch-son.

An admission

So I admit that I have never understood the Acts or the Gospels, not any literary narrative of this type. Devoted to the Greek idol and trained in Greek ideas, we understand or scorn them while only using words in the Hellenic language: allegories, parables, analogies, metaphors, symbols ..., analysis, hermeneutic ..., myths, theology ... Does one play the piano with boxing gloves?

The scene in the *Meno* where Socrates arrogantly crushes the slave boy regarding the diagonal and Euclid's hiding of the algorithm amid the *Elements* ... show how Greek thought suppressed sequences, whose tradition came from the Fertile Crescent. This tradition returned in the past through knowledge and practices (the position of arithmetic operations, algebra, infinitesimal calculus ...) and is establishing itself today through information technology. As a result, the current scorn for religions, which, in the West, generally descend from Semitic traditions, comes less from critical reasons or from atheism than from the usual blindness of one culture regarding another. Since conceptual thought agreed to consider singularity only on condition that it belong to a generality, that is to say, of destroying it, and since it understood the event only on condition of grasping it under a law, that is to say, of emptying it of its value, conceptual thought erased the specificity of algorithmic thought, which returned, formerly and recently, in the procedures of business or of operations, but is springing up again in our machines and other cognitive innovations. Lastly, uniting two gestures, erasing the disdain maintained by one culture towards its neighbouring culture, contemporary knowledge can, for the first time, think together concept and narrative, logic and literature, science and religion.

Better yet, do we attain the universal uniquely by means of the concept? On the contrary, conceptual thought reaches it poorly; when compelling itself to think the set of all sets, it runs into a paradox. Stringing operations together and advancing from events to advents, algorithmic thought constructs, on the contrary, singularities like Jesus Christ, the Abortion, Michel de Montaigne, Martin Guerre ..., in which the universal is incarnated. So, at the sight of a star throwing its rays from some point in space towards the world, everyone understands that, from every point of space, a star throws its rays towards all the other points. Touched by this light, everyone can understand without any prior abstraction.

'Others form man, I recite him,' Montaigne wrote: others format man, I excite him and recount his circumstances and advents according to changing time. Upstream, Saint Paul: the old formats plunged man into the law, the concept and jurisdiction; I suscitate him [*suscite*] in the name of the Resurrected one [*Ressuscité*].[1] Downstream and following the example of these efforts, our new cognitive branches today recite a hundred singularities of him, inciting to be reborn those singularities that sleep in the forms of information. For a long time, of man, there has been a little bit of concept and a lot of narrative; since this morning, of the things of the world, there has been a lot of concept and as much narrative; of the human and the natural, there are science and literature, stems and branches. Here is the contemporary overlapping: stable and changing, solid and fluid, format and news … narrative and idea, procedural and declarative, algorithmic and conceptual.

Will we, in the end, irreversibly return to a format? Which one? Montaigne concluded with the famous formula: 'Every man carries within himself the entire form of the human condition.' Incandescent, he loses all belongingness, from which comes all the evil in the world. Universal and empty, this form abandons all format.

Virtual contract

So no one will be mistaken: the singular example, the advent, the individual, the contingent and the news aim at the universal as much as formal formats do. Contingency's knowledge equals necessity's knowledge in dignity. Better, nothing resists the universal of the format better than the universal of the branch. Hence the exit light at the end of the tunnel of death. Do you remember zero risk, that statistical blunder? The most generous of acts, the most effective of medications … implicate their accursed portion: the law of large numbers brings catastrophe back. The gentlest of machines doesn't spare us from accident. The best of all worlds includes Evil, mixed in at Creation. That's why God himself appeared before the tribunal of the theodicy. Death is screwed fast into the secret fold [*pli*] of conceptual thought.

The equal dignity of the universal and the existential appears in the new cognition. Even if there were only one death, we would no longer tolerate it. We accept reason, but without crime; the format, but without victim. The good shepherd leaves the herd for the lost sheep. The universal of the individual completes the universal of the concept, forms a counterpart to it, offsets it, literally redeems it. We will only move towards globalization on condition of moving, at the same speed, towards the individual; reason will have to cultivate the detail of the landscape, the diversity of the living thing and the person recognized as a universal. Today, contingent existence is fighting the hominian's last battle for immortality.

Have you noticed, recently, that the latest cures adapt to each one of us? It's impossible to draw up statistics for this innocent treatment since each intervention resembles an original narrative. The scientist-doctor leaves the father-position for that of the brother: I recognize you to be singular, he says, and not to be an object of my act or an application of the cure. Thus, the intention of fabricating the human still follows the ancient projects in which only the father decided, whereas the son, now seated to his right, exists, free to do with his life as he pleases. Morality here consists in listening to the son. But how are we to give rights to all the sons still to be born?

To the scandal of many, I once proposed to accord 'nature' the status of a legal subject. Understand this term in the exact sense: things and persons to be born. Let's agree on the natural contract with daughters and sons, *naturae et naturi*, those of future generations. A complementary audacity, the same status of legal subject extends to possibles, *natura*, to things to come. Let's all answer the burning questions, on a case-by-case basis of course, but under the guarantee of this virtual contract.

Its recent return to the individual and contingency leads science to the son-position; science therefore requires, for its regulation, everyone's opinion; not deciding all by itself, in the father-position, it enters, as I have said, into a game with several players, a cognitive, ethical and sociopolitical game; inviting future generations into it wouldn't have any meaning if these possible worlds and these humans to be born didn't obtain the status of legal subjects and if we didn't sign with them, absent, a contract – a transcendental condition for knowledge and for action. Everyone sits on the tribunal that will decide these births: the public, patients, politicians, media, jurists, the religious ..., both real ones and virtual ones. The subject of knowledge and of technology becomes universalized in the concordance.

In method, the concept dialogues with the individual, as, in politics, the scientist does with the public, the father with the son and, in general, the real with the possible, the necessary with the contingent, humanity with the world. Existential, this book celebrates this new contract.

Project

On the bottom of a bark on Lake Bienne, Rousseau, solitary, felt himself existing between sky and water, among the birds and the foliage; a citizen of Geneva, he signed the *Social Contract*, at least virtually, with his peers, present or past. There is no crowd or state in nature; no flora or fauna in law. On one side, the things; humans on the other. We are continuing this perilous acosmist divorce today: history forgets geography; neither the social sciences nor politics care about the planet. Yet, not only do we inhabit the world, but today we are weaving ties with it that are so global and closely woven that it enters into our contracts.

The UN, the WHO, NATO, the FAO, UNESCO, the Red Cross, the World Bank ..., international organizations deal with our relations as though we didn't inhabit or change the Earth. Yet our conflicts tend to occupy the ground, to seize sources of water or gas wells, to appropriate seeds and species, to grant themselves the right to treat the atmosphere like a trash can ... If, like animals, we soil what we want to make our own niche, global pollution demonstrates the height – and no doubt the end – of appropriation. Let's conceive a new institution, which we could name WAFEL (Water, Air, Fire, Earth, Life), in which *Homo politicus* would welcome the elements and living things, non-appropriable quasi subjects because they form the common habitat of humanity. At the imminent risk of death, we have to bring about peace between ourselves to safeguard the world and peace with the world in order to save ourselves.

Envoi

Today, we are living through a triple branch of newnesses of a spatio-temporal and global human scope: the Grand Narrative ties history to time and to the contingent events of the Universe; amid this Narrative, the advent, once again event-oriented and contingent, of *sapiens sapiens* unifies its species, dates its origins, opens its genealogy, cadences its spread into the world and the diversification of its cultures, renders humanity family. The end of belongingness: there is no longer any North, South, East or West, rich or poor; there are only, whatever the culture they may pride themselves in, brothers-subjects, stemming from an African source, and, because of the powers they have acquired, responsible all together for humanity, for the universe and for their common evolution.

We are reaching the limits of the view projected by Saint Paul, one of the rare philosophers, to my knowledge, who had thought newness as such, the contingent event that constitutes it, the existence and the universality of the subjects brought about by its advent and the surpassing of the formats within which this new, this event-orientation and this universal became established. Let's redraw, with him, the triple stem issuing from our old Western cultures: the temporal conception of history, inherited from the writing prophets of Israel, taken over, after a Christian fashion, by Saint Augustine or Pascal, secularized by Condorcet or Auguste Comte, lastly made as erudite as you please …; the spatio-temporal view of the world issuing from

Greek science and taken over by geometers, astrophysicists or biochemists ...; community-based and solidary society, demanded lastly by every law in the world and not only Roman law, more juridical and declarative than Anglo-Saxon law, which is more algorithmic and case-based ... These three feeders in the past flowed together towards a new era; swollen, these three tributaries, history, world and society, still local, disappear and flow together today in a branch that's transcultural because connected with nature. There is no longer any history, world or society but rather the universality of space-time and of human persons.

New subjects of this conceptual and concrete universality, we put our evolution, plus the evolution of living things and of the inert world, under the beneficial or perverse feedback effect of our singular intentions and of our acts. Evolution has produced a producer of evolution.

No prophetism drives me, no proselytism impels me, yet I do not doubt that, amid the tremendous moanings of childbirth, a new era is being announced.

*   *   *

Through faithfulness to itself, this book should have ended with a story. Ever since Achilles and Sarah all the way up to Madame Bovary and Tintin, every narrative, at least those I am acquainted with, has recounted the adventures of a heroine or of a brave man, even if they related the banal circumstances lived by some anonymous person from the street. Even reversed or constructed, exceptions fascinate. Have you ever heard opera without soloists, composed solely for choirs? Have you ever read, seen or skimmed through films, television shows or comic strips without any star, whether brilliant or common? The old culture consumes personages.

It would have been necessary to write a utopian narrative, an exodus … in which, encountering some given circumstances and knowing how to negotiate them, an innumerable group would escape two combined deaths: the death one inevitably encounters and the one this group prepares with its own hands. I didn't know how to invent this narrative. I suspect that those who will conceive it will felicitously open the door that was closed before my weakness.

Notes

Preface

1   Branch = *rameau*, which is a small branch. Think of the palm branches of Palm Sunday, which is called *Rameaux*, just like this book. In this work, 'branch' will always translate *rameau*. All footnotes belong to the translator.

Format-father

1   Sol-fa notation was invented by Guido D'Arezzo, who used the first syllables of the first lines of the hymn 'Ut queant laxìs' to designate the notes (though in English, we say 'do' instead of *ut*).

2   In French, accounting is *comptabilité*, which reads as countability. So when Serres is using the term in the broad sense, I will render it as 'countability'.

3   The news = *la nouvelle*, which normally means a piece of news or news referring to some specific thing. *Les nouvelles* means news in general. But Serres seems to be playing with language here. He is clearly referring to new things or newness, but normally he would write that as *nouveauté*. Later he refers to *La Bonne Nouvelle*, Christianity's Good News, so he is referring to news in this sense as well as the newness that's opposed to the monotony of the format. I don't see much purpose in distinguishing *nouvelle* from *nouvelles* in this translation. I

merely advise the reader to keep in mind that 'news' in this work can also have strong overtones of the new.

4 Trave = *travail*, which derives from the Latin *tripalium*, a torture device having three stakes.

5 Prevision = *prévision*, which would normally be translated as prediction (as in the previous paragraph), but given the emphasis on sight in this passage, prevision seemed more apt.

Science-daughter

1 I know but don't comprehend = *je sais, mais ne connais pas. Savoir* and *connaître* both mean to know. So to try to render Serres's point into English, I have resorted to 'comprehend' for the verb *connaître* and 'comprehension' for the noun *connaissance*. But, at times in this chapter, I will also translate the latter two terms as 'understand' or 'knowledge'.

2 The word *escient* is mostly used in the phrase *à bon escient* [advisedly, wisely] in contemporary French.

3 Corneille's *Horace*, Act 3, Scene 5. This phrase uttered by the old Horatius has been called one of the most sublime in French literature. 'What did you wish that he should do against three?' Old Horatius: 'That he die.'

The adoptive son

1 Rom. 7.6.

2 'Wall, city and port, refuge from death', French readers will easily recognize the opening lines of Hugo's *Les Djinns*, except in the poem *mur* [wall] is plural.

3 A line from Ronsard's 'Elegy against the Woodcutters of Gâtine'.

4 Ferdinand Brunetière (1849–1906) was a French literary critic who wrote a book called *L'Évolution de genres dans l'histoire de la littérature* [The evolution of genres in the history of literature], in which he applied Darwinism to literature.

Event

1 Pascal famously wrote: If Cleopatra's nose had been shorter, the whole face of the earth would have been different.

2 Mark with their strikes heartbeats, coups d'état and coups de théâtre = *marque de ses coups les coups de cœur, coups d'État et coups de théâtre.* Clocks mark with their strikes or blows all the following strikes, too.

3 Sense = *sens*, which can designate either meaning or direction. I have made use of 'sense', appealing to its less obvious mathematical sense of directionality as well as the more obvious sense of meaning.

4 The first sentence of Montaigne's essay 'Of Repentance' (III 2, F 740; VS 804b).

5 *L'Arlésienne* is the name of a short story as well as a play by Alphonse Daudet. The situation described here does not appear in the short story but may possibly be found in the play.

6 To quote the cited passage from *Geometry* (London: Bloomsbury Publishing, 2017): 'Certain things traverse the sieve, others not [*pas*]: here we find not only the meaning of the word "to flow" but also that of "to pass", whose unity, in its course, is designated by the *pas* [step], when advance is positive, but that in the contrary case, when it doesn't pass, we name, not far from negation, with the *pas* of *ne … pas*. The unity of the time that *passes* must be doubled into this advancing course and this immobility frozen by some obstacle stopping the progress.' *Pas* can mean either step or the adverb not. Of course, Serres also sees *pas* in *passer*, to pass.

7   *Temps* can mean both time and weather in French. So the one kind of *temps* comes close to the other kind of *temps*.

Advent

1   In contemporary French, *encontre* mostly only exists in a few phrases meaning something akin to contrary to or against. But in Old French, it meant a meeting or encounter. Contemporary French says *rencontre*.

2   He disrupts communication because *parasite* in French also has the meaning of broadcast interference or static.

3   A First World War memoir by Jacques Péricard, but also a cry used of old to wake up soldiers in barracks.

4   To you, with you = *À toi, avec vous. Toi* is the pronoun for the second person informal. *Vous* is the pronoun for the second person formal or the second person plural.

5   Arrivals = *advenues*, a word related to *avènement*, advent. Both words signify arrival. All English forms of arrive in this passage translate a French word related to *avènement*. More generally, every instance of 'arrival' translates *avènement* or *advenue* in this work.

6   Natured and naturing refer to Spinoza's distinction of *natura naturata* and *natura naturans*.

Today

1   For 'excite' and 'suscitate', it might be best to consult their respective etymological meanings of to set in motion, and to arouse or call into life. *Ressuscité*, of course, means to call back to life, here the one called back into life.

Bibliography

Regarding scientific questions and contemporary technologies, navigating on the internet, easy, renders bibliographies superfluous. This one only mentions the basic works; works going into detail include suitable bibliographies.

System

Format-father

Brillouin, L., *Science and Information Theory*, New York: Academic Press, 1962.

Crosby, A. W., *The Measure of Reality: Quantification and Western Society (1250–1600)*, Cambridge: Cambridge University Press, 1997.

Erasmus, D., *Praise of Folly*, London: Penguin Classics, 1994.

Frémont, C., *Singularités*, Paris: Vrin, 2003.

Plato, *Statesman*, in *Plato: Complete Works*, Indianapolis: Hackett Publishing Co., Inc., 1997.

Serres, M. (director), *Éléments d'histoire des sciences*, Paris: Bordas, 1989.

Voltaire, *Histoire du docteur Akakia*, in *Mélanges*, Paris: Gallimard, coll. «Bibliothèque de la Pléiade», 1961; sur la prévoyance et la prévision, p. 295.

Science-daughter

Corneille, P., *Corneille's Horace*, Sydney: Wentworth Press, 2016.
Homer, *The Iliad*, London: Penguin Classics, 1998.
Plato, *Timaeus*, in *Plato: Complete Works*, Indianapolis: Hackett Publishing Co., Inc., 1997.
Racine, J. *Andromache*, Boston: Mariner Books, 1984.

The adoptive son

Nestle, E., ed. *Novum Testamentum Graece*, Stuttgart: Deutsche Bibelgesellschaft, 1908.
Collectif. *Traduction œcuménique de la Bible, (TOB)*, Paris: Le Cerf, 1982.
From the enormous bibliography on Saint Paul, I shall extract:
Baslez, M.-F., *Saint Paul*, Paris: Fayard, 1991.
Ben-Chorin, S., *Paulus*, Munich: Deutscher Taschenbuch Verlag, 1970.
Breton, S., *A Radical Philosophy of Saint Paul*, New York: Columbia University Press, 2011.
Lafon, G., *Épitre aux Romains; Épitre aux Galates*, Paris: Flammarion, 1987.
Martyn, J. L., *Commentary on the Galatians* (The Anchor Bible), vol. 31, New York: Doubleday, 1996.

Narrative

Event

Bergson, H., *The Creative Mind: An Introduction to Metaphysics*, New York: Citadel Press, 1946.
de Cervantes, M., *Don Quixote*, London: Penguin Classics, 2003.
de La Fontaine, J., *The Complete Fables of Jean de La Fontaine*, Champaign: University of Illinois Press, 2007.
Leibniz, G. W., *New Essays on Human Understanding*, Cambridge: Cambridge University Press, 1996.

Advent

Cailleux, A., «Premiers enseignements glaciologiques des expéditions polaires françaises», 1948, *Revue de géomorphologie dynamique*, 1952, 1, pp. 1–19.

Flaubert, G., *Madame Bovary*, New York: Bantam Classics, 1982.

Leibniz, G. W., *Theodicy*, New York: Bantam Classics, 1982.

Les Présocratiques, Paris: Gallimard, coll. «Bibliothèque de la Pléiade», 1988.

Quintilian, *De institutione oratoria*, Leipzig: Meister, 1887.

Racine, *Iphigenia, Phaedra, Athaliah*, London: Penguin Classics, 1964.

Zola, E., *Au Bonheur des Dames*, London: Penguin Classics, 2002.

Today

Euclid, *The Thirteen Books of Euclid's Elements*, translated from the text of Heiberg, ed. T. L. Heath, New York: Dover, 1956.

Montaigne, *Essays*, London: Penguin Classics, 1993.

Plato, *Meno*, Indianapolis: Hackett Publishing Co., Inc., 1980.

Rousseau J. J., *Reveries of a Solitary Walker*, Indianapolis: Hackett Publishing Co., Inc., 1992; *The Social Contract*, London: Penguin Classics, 1968.

Printed in Great Britain
by Amazon

40975188R00116